SERIES 2—NINTH EDITION

TECHNICAL DRAWING PROBLEMS

HENRY CECIL SPENCER
Late Professor Emeritus of Technical Drawing;
Formerly Director of Department
Illinois Institute of Technology

IVAN LEROY HILL
Professor Emeritus of Engineering Graphics;
Formerly Chairman of Department
Illinois Institute of Technology

JOHN THOMAS DYGDON
Professor of Engineering Graphics,
Chairman of the Department,
and Director of the Division of Academic Services
and Office of Educational Services
Illinois Institute of Technology

JAMES E. NOVAK
Associate Director/Executive Officer,
Office of Educational Services
Illinois Institute of Technology

Macmillan Publishing Company
NEW YORK

Maxwell Macmillan Canada
TORONTO

Maxwell Macmillan International
NEW YORK   OXFORD   SINGAPORE   SYDNEY

Macmillan Publishing Company
866 Third Avenue, New York, New York 10022

Macmillan Publishing Company is part of the Maxwell Communication Group of Companies.

Maxwell Macmillan Canada, Inc.
1200 Eglinton Avenue East
Suite 200
Don Mills, Ontario M3C 3N1

ISBN 0-02-414670-6

Printing: 1 2 3 4 5 6 7 8   Year: 1 2 3 4 5 6 7 8 9 0

Preface

Technical Drawing Problems, Series 2, is intended primarily for use with *Technical Drawing* (Ninth Edition, 1991) by Giesecke, Mitchell, Spencer, Hill, Dygdon, and Novak, published by Macmillan Publishing Company. All references and instructions refer to that text. However, this workbook may be used with any good reference text.

Since the time available for the teaching of technical drawing is limited, the objective has been to create a collection of problem sheets that eliminate repetitious drawing and still give adequate coverage of the fundamentals. It is expected that in many cases the instructor will supplement these problem sheets with assignments of problems from the text to be drawn on blank paper.

Most of the problems in this workbook are taken directly from industry, and are carefully selected from thousands of original drawings in drafting department files. Although a given print from industry may in itself be too simple or too complicated for classroom use, it is useful as a basis for the classroom problem. There is no problem in drawing that does not have its counterpart on prints used in industry. By means of many notes, supplementary drawings, and pictorials, every effort has been made to connect the problems to the practical situations from which they were taken.

An outstanding feature of this revision is that fractional-inch dimensions have largely been replaced by the decimal-inch and metric dimensions now used extensively in industry. A decimal and millimeter equivalents table and appropriate full-size and half-size scales are provided inside the front and back covers for the student's convenience.

Technical sketching is recognized as an important skill in engineering work, and an entire unit on multiview and isometric sketching is included early in the book. Grids are provided on the first sheets to help the student gain confidence. The instructor may assign many of the problems to be drawn freehand for additional emphasis of this technique.

All sheets are 8.5" × 11.0" in conformity with American National Drafting Standards for engineering drawing practices, a size that facilitates handling and filing by the student and the instructor. Several problems are printed on vellum to provide experience in the manner of commercial practice. It is expected that in most cases the instructor will supplement these problem sheets with assignments of problems from the text to be drawn on blank paper, vellum or film.

All of the problems are based upon actual industrial designs, and their presentations are in accord with the latest ANSI Y14 American National Standard Drafting Manual and other relevant ANSI standards.

In response to the increased usage of computer technology for drafting and design, a number of problem sheets in computer-aided drafting (CAD) have been included. The problem sheets on detail drawings are presented to provide practice in making regular working drawings of the type used in industry. These are suggested for solution either by a computer-aided drafting system or by traditional drafting methods.

A special feature of the workbook is the Instructions, in which a detailed explanation of the requirements of each problem is given. together with specific references in the text to cover each point. These instructions are expected to anticipate many of the questions that will arise in the student's mind in connection with each problem, thereby freeing the instructor from much detailed individual instruction and providing more time for aid to students who have basic difficulties.

Most of the problems have been used extensively in engineering colleges throughout the country and, therefore, are well tested under classroom conditions. The cooperation of many schools and companies in supplying the original prints for these problems is deeply appreciated. The valuable contributions of Prof. H. E. Grant, originally a co-author of this book, are gratefully acknowledged.

Comments and criticisms from the users of this workbook will be most welcome.

Ivan Leroy Hill
Clearwater, FL

John Thomas Dygdon
*Illinois Institute
of Technology
Chicago, IL*

James E. Novak
*Illinois Institute
of Technology
Chicago, IL*

cooperating companies

Allis-Chalmers Manufacturing Company
American Brake Shoe and Foundry Co.
American Can Company
American Laundry Machinery Company
American Monorail Company
American Tool Works Company
Arter Grinding Machine Company
Atlas Press Company
Avey Drilling Machine Company

Baker Brothers, Inc.
Baldwin Locomotive Company
Barber-Colman Company
Barrett, Haentjens and Company
Barry-Wehmiller Mach. Company
Bauer Brothers Company
Black-Clawson Company
Blanchard Machine Company
Bodine Corporation
Boye and Emmes Machine Tool Company
Bridgeport Machines, Inc.
Buffalo Forge Company
Buffalo Foundry and Mach. Company
Bullard Company, The
Busch-Sulzer Diesel Engine Co.
Butterworth, H. W., & Sons Company

Caterpillar Tractor Company
Chance-Vaught Aircraft Division
Cincinnati Bickford Tool Company
Cincinnati Milling Machine Co.
Cincinnati Mine and Machine Company
Cincinnati Planer Company
Cincinnati Shaper Company
Clark Machine Company
Cleveland Automatic Machine Company
Cleveland Tractor Company
Cleveland Twist Drill Company
Consolidated Mach. Tool Company
Consolidated Packaging Machine Corp.
Crane Company
Curtis Mfg. Co.

Davis and Thompson Company
Davenport Machine Tool Company
Day, J. H., Company
Defiance Machine Works
Drever Co.
Dumore Company

Eastern Engineering Company
Euclid Road Machinery Company

Fairbanks-Morse Company
Farnham Mfg. Company
Farrell-Birmingham Company
Federal Machine & Welder Company
Footburt Company
Fosdick Machine Company
Frigidaire

Gallmeyer and Livingston Company
Gardner Machine Company
Gairing Tool Company
General Engineering Company
General Machine Corporation
General Railway Signal Company
Giddings-Lewis Mach. Tool Company
Gisholt Machine Co.
Gorton, Geo., Company
Goss and DeLeeuw Mach. Company
Gray, G. A., Company

Greenfield Tap and Die Company
Greenlee Brothers and Company

Hamilton Tool Company
Hammond Machinery Builders, Inc.
Hanson Whitney Machine Company
Haskins, R. G., Company
Henry and Wright Mfg. Company
Hill Acme Company
Hudson-Sharp Machine Company
Hunter Engineering Company
Hydraulic Press Mfg. Company

Illinois Tool Works
Ingersoll Milling Mach. Company

Kearney and Trecker Corp.
Kelley Reamer Company
Kiefer, Karl, Mach. Company
Koehring Company

Lamson and Sessions Company
Landis Machine Company
Lees-Bradner Company
Lehmann Mach. Company
Lester Eng. Company
Liberty Planers, Inc.
Lincoln Electric Company
Link-Belt Company
Liquid Carbonic Company
Lodge and Shipley Company

McCulloch Engineering Corporation
Madison Kipp Company
Martin, Glenn L., Company
Mattison Mach. Works
Medart Company, The
Merrow Machine Company
Miehle Printing Press and Mfg. Company
Milwaukee Foundry Equipment Company
Monarch Governor Company
Monarch Mach. Tool Company
Murphy Diesel Company

National Supply Company
New Britain-Gridley Machine Company
New Departure Bearing Company
Niagara Machine and Tool Works
Niagara Mach. Tool Company
Nordberg Manufacturing Company
Northwest Engineering Company
Norton Company

Oilgear Company

Package Machinery Corporation
Polaroid Corporation
Pratt and Whitney Aircraft Company
Proctor and Schwartz, Inc.

Racine Tool and Machine Company
Reed Prentice Corporation
Rivett Lathe and Grinder, Inc.
Rock Island Arsenal
Rockford Machine Tool Company

Sellers, William, & Company
Sheffield Corporation
Sidney Machine Tool Company
S. K. F. Industries, Inc.
Skinner Chuck Company
Smith Engineering Works
Socony-Vacuum Oil Company
Springfield Arsenal

Springfield Mach. Tool Company
Stephan Nat'l. Industrial Advertising
Stokes, F. J., Machine Company
Stokes and Smith Company
Sun Manufacturing Company
Sunstrand Machine Tool Company
Superior Engine Division

Taylor Manufacturing Company
Templeton, Kenly & Co.
Thompson Products, Inc.

Union Manufacturing Company
United States Automatic Box Mach. Co.
United States Department of Interior
Universal Power Shovel Corporation

Van Norman Machine Tool Company
Vilter Manufacturing Company

Waltham Machine Works
Waltham Watch Company
Warner and Swasey Company
Watertown Arsenal
Weldon Tool Company
Western Austin Company
Western Electric Company

Yoder Company

foreign companies and schools

Ateliers des Charmilles S.A., Geneva, Switzerland
Benese, Dra. E., Vysoke Skoly Technicke, Brne,
    Czechoslovakia
Bertram, John, and Sons Company, Ltd.,
    Dundas, Ontario, Canada
B. & S. Massey Ltd., Manchester, England
Brown-Boggs Foundry and Machine Company,
    Hamilton, Ontario, Canada
Canada Machinery Corp., Galt, Ontario, Canada
Delamere and Williams Ltd., Toronto, Canada
Ecole Nationale Professionnelle, Creil, France
Ecole Nationale Professionnelle, Metz, France
Ecole Nationale Professionnelle, Nantes, France
Hepburn, John T., Ltd., Toronto, Canada
Herbert, Alfred, Ltd., Coventry, England
Ing. Santiago A. Cerna, Fabricas Monterrey S.A.,
    Monterrey, Mexico
Lang, John, and Son, Ltd., Johnstone, Scotland
McDougall, R., Company, Ltd., Galt, Ontario,
    Canada
Modern Tool Works, Toronto, Canada
Parkinson, J., & Sons, Shipley, England
Skoda Works, Plzen, Czechoslovakia
Stewart, Duncan and Company, Ltd., Glasgow,
    Scotland
Suomen Sahko O. Y. Gottfred Stromberg,
    Helsinki, Finland
Svenska Turbinfabriks A-B, Ljungstrom,
    Finspong, Sweden
Taylor and Hubbard Ltd., Leicester, England
Taylor, Chas., Ltd., Birmingham, England
Technical University, Helsinki, Finland
Techniska Laroverket, Helsinki, Finland
Tullis, D. J., Ltd., Clydesbank, Scotland
Usines Carels Freres, Ghent, Belgium
Ward, H. W., and Company Ltd., Birmingham,
    England
Woodhouse and Mitchell Ltd., Brighouse, York,
    England
Yates, P. B., Machine Company, Hamilton, Canada

Contents

Worksheets

Groups	Sheets	Groups	Sheets
A Instrumental Drawing	A-1 to A-4	L Revolution	L-1 to L-2
B Technical Lettering	B-1 to B-8	M Isometric Drawing	M-1 to M-4
C Technical Sketching	C-1 to C-4	N Oblique Projection	N-1 to N-3
D Geometric Constructions	D-1 to D-3	O Threads and Fasteners	O-1 to O-7
E Normal and Inclined Surfaces	E-1 to E-3	P Springs	P-1
F Inclined and Oblique Surfaces	F-1 to F-4	Q Dimensioning	Q-1 to Q-7
G Cylindrical Surfaces and Combinations	G-1 to G-3	R Detail Drawings	R-1 to R-11
		S Design Layouts	S-1 to S-2
H Intersections	H-1 to H-2	T Charts and Graphs	T-1 to T-2
J Sections	J-1 to J-10	U Graphical Mathematics	U-1 to U-5
K Auxiliary Views	K-1 to K-7	V Computer-Aided Drafting	V-1 to V-5

Vellums

The following sheets, printed on vellum, are located at the back of this workbook, after Sheet V-5.

Alphabet of Lines	A-4	Auxiliary Section	K-6
Normal Surfaces	E-1	Isometric-Irregular Combinations	M-3
Inclined Surfaces	E-3	Square Threads	O-3
Combination Edges and Surfaces	G-3	Fasteners	O-7
Full Section Views	J-2	Springs	P-1
Broken-Out and Removed Sections	J-9	Limit Dimensioning	Q-4
Aligned Sections	J-10		

Instructions

References are to Ninth Edition of *Technical Drawing (1991)*
by Giesecke, Mitchell, Spencer, Hill, Dygdon, and Novak.

Throughout this workbook alternative dimensions, often not the *exact* equivalents, are given in millimeters and inches. Although it is understood that 25.4 mm = 1.00", it is more practical to use approximate equivalents: 25 mm for 1.00", 12.5 or 12 mm for .50", 6 mm for .25", 3 mm for .12", etc. Exact equivalents should be used when accurate fit or critical strength is involved.

GROUP A. INSTRUMENTAL DRAWING

A-1. Drawing Equipment. References: §§2.1–2.10, 2.16, 2.24–2.30, 2.33–2.44, 2.48, 2.49, 2.53, 2.56. Before starting any sheet, letter your name and file number in the title strip. Draw extremely light horizontal guide lines from the starting marks shown, using a sharp 4H lead, then draw light vertical (or inclined—see instructor) guide lines at random, and letter with a sharp F lead, giving your last name first. Numbers in the column at the right correspond to the encircled numbers on the drawing. Draw light vertical (or inclined) guide lines and letter the correct names of the corresponding items of equipment or brief answers to the questions. Use abbreviations where necessary. Items 1, 2, and 3 refer to *parts* of the T-square; items 6 and 7 refer to the correct name of the triangle; item 8 refers to the *number of degrees;* item 13 refers to the length of the extended portion of the sharpened lead; items 19, 21, and 23 require an answer of "yes" (make the test); items 20, 22, 24, and 25 refer to the amount of error measured to the nearest 0.4 mm or .02" (1/64"); and items 46 to 50 require the correct reading of the scales shown.

A-2. Drawing Lines. References: §§2.7–2.17, 2.25–2.27, 2.29, 6.1–6.7.

Spaces 3, 5, 7. Study §§6.2–6.7. Carefully draw freehand sketches as indicated, using a sharp F lead and making triangles about the same size and shape as those shown in *Space 1.*

Draw lines to fill spaces as follows.

Space 2. Draw horizontal lines 12 mm (or .50") apart. Use the full-size scale, Fig. 2.35 (a) (or Fig. 2.36 (a)), in the position shown in Fig. 2.54 (III) and set off 12 mm (.50") distances with a sharp 4H lead, Figs. 2.9 and 2.11 (c). Then, with a sharp F lead, draw horizontal **black** visible lines, Fig. 2.14, as shown in Fig. 2.20 (a).

Space 4. Draw vertical hidden lines 12 mm (.50") apart. Use the full-size scale as shown in Fig. 2.54 (II) and set off 12 mm (.50") spaces as shown in Fig. 2.38 (a). Then draw vertical hidden lines upward, Fig. 2.14, as shown in Fig. 2.21 (a). Make dashes 3 mm (.12") long and spaces 0.8 mm (.03"). Actually measure the first line of dashes; then space the others by eye. *Accent the ends of each dash,* Fig. 2.8 (c). Use a sharp F lead, and make lines **black.**

Space 6. Draw construction lines 45° with horizontal upward to right 12 mm (.50") apart. These lines will slope upward to the right at 45° as shown by the triangle in Fig. 2.23 K. Therefore, distances between lines must be set off at right angles to this. First find the center of the space by means of light construction diagonals crossing at the center. Through this center draw, using a sharp 4H lead and with a 45° triangle in the position shown in Fig. 2.23 D, a construction line at right angles to the required lines. Set off along this line the 12 mm (.50") spaces, and through these draw extremely light construction lines (4H lead) as required. Hold your sheet at arm's length. If you see these lines plainly, they are probably too dark, §2.46.

Space 8. Draw visible lines 60° with horizontal downward to right 12 mm (.50") apart. First find the center as in *Space 6;* then draw a construction line that will be at right angles to the required lines. For this your triangle should be in the position shown in Fig. 2.23 L. On this construction line, set off the 12 mm (.50") spaces, and draw the required visible lines through these points with your triangle in the position shown in Fig. 2.23 E. Use a sharp F lead and make lines **black.**

A-3. Drawing Lines. References: §§2.7–2.17, 2.21, 2.22, 2.25–2.27, 6.1–6.7.

Spaces 1, 3 5, and 7. Draw carefully *freehand* sketches as indicated, using a sharp F lead and sketching the triangles and T-squares about the size of those shown in *Space 1.*

Draw lines to fill spaces as follows.

Space 2. Draw cutting-plane lines 15° with horizontal upward to right 12 mm (.50") apart. Set your triangles as in Fig. 2.23 R so that a line along the upper side of the 45° triangle will be sloping upward to the right as required in this problem. The other leg of the 45° triangle will then be at right angles to this, and the diagonal construction line through the center of the rectangle can be drawn along it, with measurements at right angles to the required lines. Cutting-plane lines are used in sectioning, Fig. 9.8. Make the dashes as shown in Fig. 2.14, the same weight as visible lines. Use a fairly sharp F lead, and make lines **black**.

Space 4. Draw center lines (℄) 75° with horizontal upward to right 12 mm (.50") apart. Set your triangles as in Fig. 2.23 T. The upper leg of the 45° triangle will then be at right angles to the required lines, and the diagonal construction line through the center of the rectangle can be drawn along it, with measurements at right angles to the required lines. Use a 2H lead, sharpened as in Fig. 2.11 (c), and make your lines *sharp*, but **black**, Fig. 2.14. Center lines are used on drawings to indicate symmetry, Fig. 6.44.

Space 6. Draw extension lines parallel to given line through given points. Follow §2.21, using a sharp 2H lead, Fig. 2.11 (c). Make lines extremely *sharp*, but **black**. Extension lines, Fig.2.14, are used in dimensioning, Fig. 13.2 (a).

Space 8. Draw section lines perpendicular to given line through given points. Follow §2.22, using a sharp 2H lead, Fig. 2.11 (c). Make lines extremely *sharp*, but **black**. Section lines, Fig. 2.14, are used for section lining, §9.4.

A-4. Alphabet of Lines. References: §§2.11–2.13, 2.33–2.37, 2.39, 2.40, 2.42–2.47. In order to learn the correct technique of drawing all the common lines in technical drawing, and the proper order of penciling a drawing, you are to make an exact copy of the drawing shown, drawing the circular view at center **A**, and omitting all instructional notes shown in *inclined* lettering. The only lettering that you are to copy is the 200 mm dimension and the finish mark. On your drawing, carefully use all measurements indicated, such as size of arrowheads, height of dimension numerals, etc. Do not give dimensions for these on your drawing.

First. Draw center lines, Fig. 7.6 (II), using a sharp 2H lead. Block in the two views with extremely light construction lines, Fig. 7.6 (III), using a sharp 4H lead. For the arcs, use your bow pencil with a 4H lead sharpened as in Figs. 2.43 (b) and 2.52 (b). Adjust needle point with shoulder end out as in Fig. 2.52 (b). Use your bow dividers to take arc radii from the given drawing, and then draw the arcs *lightly* from center A below. Next, draw straight horizontal and vertical construction lines in the circular view and then in the right-hand view. Note that horizontal lines can be projected across from view to view and that vertical lines can be projected down from the drawing above.

Second. Heavy in the views and finish the drawing, using a sharp F lead in your bow pencil for the arcs, sharpened as in Fig. 2.43 (b). All final lines should be *clean* and **dark**. Use both hands on the bow pencil for all circular hidden dashes in order to control the lengths of the dashes. Use the large compass only for arcs of more than 25 mm (1.00") radius.

Third. Draw extension lines and dimension line **dark** but *very sharp*, using a 2H lead. Draw horizontal and vertical guide lines *lightly* for dimension numerals with a 4H lead, Fig.4.27 (b). For arrowheads, see Fig. 13.7. *Note:* Do not erase construction lines. They should be so light that they are barely visible.

GROUP B. TECHNICAL LETTERING

B-1. Vertical Capitals. References: §§2.12, 2.13, 4.1–4.18, 4.24. Fill in the spaces with freehand letters as indicated. For the large letters in Exercise 1, use a fairly sharp HB lead; for the rest of the sheet use a sharp F lead. Draw both horizontal and vertical guide lines as indicated in Exs. 3 and 4, using a sharp 4H lead. Your lead should be fairly sharp for the larger letters and *very sharp* for the smaller letters. All letters should be *clean-cut* and **dark**. The larger letters may be sketched lightly first, erased and corrected as necessary, but finally drawn with firm, *clean*, **dark** strokes. The smaller letters should be lettered directly with single strokes.

If you are left-handed, you need not use the strokes shown. Instead use the strokes that seem to give the best results, §4.13. However, you must learn the correct proportions of the letters. On all lettering sheets, add a note that you are left-handed.

B-2. Vertical Capitals and Numerals. References and instructions same as for B-1.

B-3. Vertical Numerals. References and instructions same as for B-1. In addition, study §§4.20, 4.26, 4.27, 13.5, 13.7, 13.11, 13.14, 13.17. For B-3b, see Fig. 4.44, and make your title perfectly symmetrical about the center line, using the method shown in Fig. 4.36 (b). For the lower half of B-3b, draw complete guide lines for all characters, using a 4H lead. Use a sharp F lead for lettering and arrowheads, and draw dimension lines with a sharp 2H lead. For the angles, refer to Fig. 13.17. For the arrowheads, see Fig. 13.7. For the small dimensions, see Fig. 13.11. For the radii, see Fig. 13.19. Use a compass for the dimension arcs.

B-4. Vertical Lowercase Letters. References: §§ 4.1–4.17, 4.21, 4.22, 4.24. Fill in the spaces with freehand letters as indicated. For the large letters in Exs. 1, use a sharp HB lead, and for the rest of the sheet use a sharp F lead. Draw vertical guide lines, as indicated in Exs. 3 and 4, using a sharp 4H lead. Your lead should be fairly sharp for the larger letters, and *very sharp* for the smaller letters. All letters should be *clean-cut* and **dark**. The larger letters may be sketched lightly first, erased and corrected as necessary, but finally drawn with firm, *clean,* **dark** strokes. The smaller letters should be lettered directly with single strokes, as indicated. Use vertical capitals and lowercase letters for your name in the title strip.

B-5. Inclined Capitals. References: §§2.14, 4.1–4.17, 4.19, 4.24. Fill in the spaces with freehand letters as indicated. For the large letters in Ex. 1, use a sharp HB lead, and for the rest of the sheet use a sharp F lead. Draw both horizontal and inclined guide lines as indicated in Exs. 3 and 4, using a sharp 4H lead. Your lead should be fairly sharp for the larger letters, and *very sharp* for the smaller letters. All letters should be *clean-cut* and **dark**. The larger letters may be sketched lightly first, erased and corrected as necessary, but finally drawn with firm, *clean*, **dark** strokes. The smaller letters should be lettered directly with single strokes, as indicated.

If you are left-handed, you need not use the strokes shown. Instead, use the strokes that seem to give the best results, §4.13. However, you must learn the correct proportions of the letters. On all lettering sheets, add a note that you are left-handed.

B-6. Inclined Capitals and Numerals. References and instructions same as B-5.

B-7. Inclined Numerals. References and instructions same as for B-5. In addition, study §§4.20. 4.26, 4.27, 13.5, 13.7, 13.11, 13.14, 13.17. For B-7b, see Fig. 4.44, and make your title perfectly symmetrical about the center line, using the method shown in Fig. 4.36 (b). For the lower half of B-7b, draw complete guide lines for all characters, using a 4H lead. Use a sharp F lead for lettering and arrowheads, and draw dimension lines with a sharp 2H lead. For the angles, refer to Fig. 13.17. For the arrowheads, see Fig. 13.7. For the small dimensions, see Fig. 13.11. For the radii, see Fig. 13.19, Use a compass for the dimension arcs.

B-8. Inclined Lowercase Letters. References: §§4.1–4.17, 4.21, 4.23, 4.24. Fill in the spaces with freehand letters, as indicated. For the large letters in Ex. 1, use a sharp HB lead, and for the rest of the sheet use a sharp F lead. Draw inclined guide lines, as indicated in Exs. 3 and 4, using a sharp 4H lead. Your HB lead should be fairly sharp for the larger letters, and *very sharp* for the smaller letters. All letters should be *clean-cut* and **dark**. The large letters may be sketched lightly first, erased and corrected as necessary, but finally drawn with firm, clean, dark strokes. The smaller letters should be lettered directly with single strokes, as indicated. Use capital letters and lowercase letters for your name in the title strip.

GROUP C. TECHNICAL SKETCHING

C-1. The "Glass Box." References: §§6.1–6.8, 6.11–6.13, 6.18–6.22, 6.25, 6.26, 6.29–6.32, 7.1, 7.2. Use a fairly sharp HB lead and follow instructions on the sheet. Sketch all six views. Include all hidden lines on the six views, but omit them on the isometric. Study §6.22, and letter in the upper right portion of the sheet the views you regard as necessary for the clearest shape description. Start the isometric sketch at point A, as indicated. Note that the isometric position is not the same as in the glass box illustration on the sheet.

C-2. Pictorial Representation. References: §§6.1–6.8, 6.11–6.13, 6.18–6.22, 6.25, 6.26, 6.29–6.32, 7.1, 7.2, 7.12–7.14. Each problem is composed of three views and an isometric. Study the given views and then complete the drawings freehand. To obtain sizes, count the corresponding grid spaces in the multiview drawings and the isometrics. Omit hidden lines in the isometrics, §18.10. Use an HB lead and make lines **dark.**

C-3. Freehand Sketching. References and instructions same as for C-2. Using a fairly sharp HB lead, sketch the six views of the object whose front and right-side views are given completely. Make lines *clean-cut* and **black.** Try to obtain the technique shown on the given lines—that is,

get *freedom* and *variety* in your lines. Complete the freehand sketch of the isometric pictorial, omitting hidden lines. List the minimum necessary views that best describe the shape of the object. For sketching the ellipses, see §16.13.

C-4. Freehand Sketching. References: §§6.3–6.10, 6.18–6.32. Using a fairly sharp HB lead, sketch the required views of the object shown. The locations of the views are indicated by corner marks. Do not try to scale the pictorial drawing, but estimate distances and proportions as explained in §6.10. A paper "scale" may be used if desired. Block in views lightly and obtain instructor's OK before darkening the lines. You will be graded largely on how well you maintain the correct proportions. *No mechanical aids are permitted* except for guide lines for the lettering in the title strip. Your lines must be *clean-cut* and **dark** and exhibit *good freehand technique,* §6.5.

GROUP D. GEOMETRIC CONSTRUCTIONS

D-1. Geometric Constructions. With each problem is shown a pictorial drawing of a practical application of the construction required. Arrows indicate the direction of sight for the views to be completed by your constructions. Use a sharp 4H lead *lightly* for all construction lines (do not erase), and a sharp F lead for all required lines (as heavy and **dark** as the given visible lines).
Prob. 1. Reference: Fig. 5.23 (d).
Prob. 2. Reference: Fig. 5.28 (a).
Prob. 3. Reference: Fig. 5.24 (a).
Prob. 4. Reference: Fig. 5.26 (c).
Prob. 5. Reference: Fig. 5.56. Draw only the visible ellipse.
Prob. 6. References: §§2.54, 5.51. Be careful to draw a smooth, symmetrical ellipse.

D-2. Geometric Constructions. Draw all required lines with a sharp F lead, and all construction lines *lightly* with a sharp 4H lead (do not erase).
Prob. 1. References: Figs. 5.10, 5.11.
Prob. 2. Reference: Fig. 5.8.
Prob. 3. References: Fig. 5.15. See also Fig. 15.2 (a).
Prob. 4. References: §§2.17, 2.43.
Prob. 5. References: §§2.18, 2.43, 5.33, 12.20. Note that the scale is half-size. Use protractor to locate holes.
Prob. 6. Reference: Fig. 5.36 (b).

D-3. Geometric Constructions. Draw all construction lines *lightly* with a sharp 4H lead, and do not erase. Draw all required lines with a sharp F lead, **dark** and *clean* to match the given lines. *Show all points of tangency* by means of light construction lines as shown in Prob. 2. Fill in the sentences asking for the *total* number of tangent points in each problem, except that in Prob. 2 do not count corners of hub as tangent points, and in Prob. 3 do not count the four small fillets and rounds.
Prob. 1. Reference: Fig. 5.35 (a).
Prob. 2. References: Figs. 5.34 (b), 5.36 (c).
Prob. 3. References: Figs. 5.37, 5.35 (b).
Prob. 4. Reference: Fig. 5.41.
Prob. 5. References: §§5.39–5.41.

3

GROUP E. NORMAL AND INCLINED SURFACES

E-1. Normal Surfaces. References: §§2.46. 2.47, 2.59, 6.25, 6.26, 6.30, 6.31, 7.1–7.3, 7.5, 7.7, 7.19, 7.20. Follow instructions on the sheet. Omit dimensions unless assigned. Study the illustration at the right of the isometric and be able to explain the function of the Necking Tool Post. If your lines are not **black**, a clear print cannot be made from your drawing.

E-2. Normal and Inclined Surfaces. References: §§6.18, 6.19, 6.25, 6.28, 6.31, 7.12–7.22. Use a *sharp* F lead, and sketch freehand or draw, as assigned, the missing lines in each problem. Make your lines **dark** and *clean*, and be careful to make hidden-line dashes about 3 mm (.12") long and 0.8 mm (.03") apart, and to join them properly. If you get "stuck" on a problem, carve out a soap model, make a clay or wood model, or make an isometric sketch, §6.12.

E-3. Inclined Surfaces. References: §§2.46, 2.47, 6.25, 6.26, 7.17, 7.21, 7.22, 7.33. Follow instructions on the sheet and draw the required views. Use a 4H lead for construction lines (extremely *light*—do not erase) and a sharp F lead for visible lines and hidden lines. Add numbers to the front view as indicated, making them small to match those in the side view, §7.6. Note that the numbered inclined surface has the same number of edges in the side and front views, §7.17. The dimension 10 × 7 indicates that the groove is 10 mm wide and 7 mm deep. Use dividers for projecting from the side view to the top view, Fig. 7.7 (a).

GROUP F. INCLINED AND OBLIQUE SURFACES

F-1. Inclined Surfaces and Oblique Edges. References: §§7.17, 7.18, 7.21, 7.22, 7.24. Draw the required views, using a sharp F lead. Fill in the table, using guide lines and making numerals the same size as those given. List the numbers in clockwise order, starting each series with the upper left number in the view.

F-2. Inclined Surfaces and Oblique Edges. References: §§7.17, 7.18, 7.21, 7.22, 7.24. Draw right-side views, using a sharp F lead. Fill in the table, using guide lines and making the numerals the same size as those given. List the numbers in clockwise order, starting each series with the upper left number in the view. In addition, in Probs. 2 and 3, number the surfaces in the side views as indicated. Make your numbers .06" (1/16") high.

F-3. Oblique Surfaces. References: §§7.17, 7.18, 7.23–7.25. In each space draw or complete the required third view, using a sharp F lead. Study the illustration at the top of Prob. 2, and using a piece of cardboard as a model, try the different positions shown. In Prob. 3, count the surfaces and edges as indicated, and letter the numbers neatly .12" (1/8") high in the spaces provided. A convenient method of counting edges is illustrated in Fig. A. It is suggested that you use the top view in Prob. 3 for counting of edges, imagining it to be the entire object. See §7.8.

(5 picas here for ART from negative)

Fig. A. Method of Counting Edges

F-4. Inclined and Oblique Surfaces. References: §§7.17, 7.18, 7.23–7.25. In each problem the top view is to be completed. Use a sharp F lead, and make your lines **dark** and sharp. Fill in the blanks in the statements adjacent to each problem. A surface is a single plane bounded all around by lines. Also fill in the table in Prob. 1, using guide lines and making numbers the same size as those given. List the numbers in clockwise order, starting each series with the upper left number in the view.

GROUP G. CYLINDRICAL SURFACES AND COMBINATIONS

G-1. Cylindrical Surfaces. References: §§6.25, 6.26, 6.30, 6.31, 7.27–7.29. Study especially Figs. 7.32 and 7.33. Use a sharp F lead and sketch freehand or draw, as assigned, the missing lines. Transfer depth dimensions from top to side views with dividers. Make your lines the same weight as those given. In Spaces 1 and 2, fill in answers to questions.

G-2. Cylindrical Surfaces. References: §§6.25, 6.26, 7.9, 7.27–7.29. Draw required views, using a sharp F lead, and include all hidden lines. Be sure to add center lines where needed. In Prob. 5, only a half-view, Fig. 7.11 (c), is required. Do not convert the center line into a visible line. In Probs. 2, 3, and 5, fill in the blanks of the statements. In all problems, transfer depth dimensions from top view to side view with dividers. In counting cylindrical surfaces, a surface is regarded as one that forms a closed figure. Two surfaces may line up to form a single circle or arc in another view.

G-3. Combination Edges and Surfaces. References: §§2.13, 2.46, 2.47, 7.5–7.7, 7.34–7.36, 13.17. Draw with instruments to the scale indicated the top and side views, showing all hidden lines and omitting dimensions. Make fillets and rounds very carefully to match those shown in the front view. Use the bow pencil, sharpened and carefully adjusted as shown in Figs. 2.43 (b) and 2.52 (b), for drawing the larger fillets and rounds. Draw the smaller ones carefully freehand to match those given. Use a sharp F lead, and make all lines *clean* and **dark** so that a clear print can be made from your drawing. Add necessary finish marks.

GROUP H. INTERSECTIONS

H-1. Cylindrical Intersections. References: §§ 2.53, 7.32, 21.27. Plot enough points on each curve so that they are spaced at intervals of approximately 3 mm (.12") along the curve. Sketch the curve very lightly freehand through the points before using the irregular curve. These curves must be symmetrical about their center line. Show all hidden lines.

H-2. Plane Intersections. References: §§2.53, 5.47, 7.27–7.33, 21.3, 21.7, 21.11, 21.21, 21.22, 21.30. Use a sharp 4H lead for construction lines, and a sharp F lead for required lines.

Prob. 1. To draw the left end of the top view, extend the sides of the cone in the front view to the vertex on the center line extended; then project to the top view and draw the conical outline. To draw the curve of intersection formed by the horizontal plane, imagine each dotted vertical line at the left end of the front view to be the edge of a plane, cutting circles on the cone of various diameters. Project across from points 1, 2, 3, 4, . . . , to the vertical center line in the right-side view; then draw circles with these radii in the side view. It can easily be seen where each circle pierces the horizontal plane of the cut at the top of the side view. Transfer these points to the top view with dividers.

The V-notch is produced by two planes that intersect at the bottom of the notch. The method is shown for projecting points to the top view. These points are the piercing points of elements of the cylinder in the plane surface, Fig. 7.31. Use irregular curve, §2.54, to draw the final curves, but sketch light lines through the points first.

At the head of each leader, letter .12" (1/8") high (use No. 4 setting of Ames Lettering Guide, and alternate holes in center column) the names of the geometric surfaces or plane curves. These names should be in accord with the actual shapes in space, and not necessarily as they *appear* in the views. Thus, an ellipse that is projected as a circle should be designated an ellipse.

Prob. 2. Use the same method for drawing the curve resulting from the intersection of plane and cone as in Prob. 1. If a sphere is intersected by a plane, the curve is always a circle. At the head of each leader, letter the names of the geometric surfaces or plane curves.

GROUP J. SECTIONS

J-1. Full and Half Sections. References: §§6.5–6.7, 6.28, 9.1–9.7. Use a fairly sharp HB lead for visible lines, and a sharp F lead for section lines spaced about 2.5 mm (.10") apart and extremely fine. Make all lines **black**. In Prob. 3, draw hidden lines in the unsectioned half of the half section.

J-2. Full-Section Views. References: §§7.33, 9.1–9.6, 12.3–12.5, 12.20. Draw full sections, using a sharp F lead for visible lines, and a sharp 2H lead for section lines and center lines. Omit all hidden lines. Make fillets and rounds carefully with the bow pencil to match those in the given views. Space section lines about 4 mm (.16") apart. Add finish marks to sectioned views, §13.17. In a sectioned area, omit the section lines where the finish mark is placed. Drilled holes, counterbored holes, reamed holes, etc., are understood to be finished; hence, finish marks are omitted.

J-3. Full and Half Sections. References: §§6.5, 9.1–9.7. In each full-section problem, the section-lined portions of the sectional views are shown, but many visible lines behind the cutting planes have been omitted. Add all missing lines freehand or mechanically, as assigned. All sections are understood to be full sections except where half sections are specified. In each half-section problem, complete the sectional view, adding all hidden lines in the unsectioned half. Using a sharp 2H lead, add all necessary center lines, §6.26.

J-4. Full and Half Sections. References: §§6.5, 7.34, 9.1–9.7, 12.5, 13.17. Draw sectional views freehand or mechanically as assigned. All are full sections except Probs. 4 and 6, which are to be half sections. For mechanical drawing, use a fairly sharp F lead for visible lines and a sharp 2H lead for section lines and center lines. Add finish marks in the sectional view of Prob. 1. All other problems are finished all over, and on a complete working drawing the note FAO would be given. Note the CORE in Prob. 1, which means that the cavity in the casting is formed by a core, and the surfaces inside are therefore all *rough*. The small hole at the right end is not a threaded hole.

In Prob. 2, note how the bottom of the drilled hole is drawn, Fig. 7.40 (a). In Prob. 3, the counterbored hole is formed as shown in Fig. 12.19 (e). Although the cutting plane is usually omitted in such cases as these, where the location of the cutting plane should be obvious, add them to all six problems to learn the correct representation, §9.5 and Fig. 9.13 (b). Cutting-plane lines take precedence over center lines.

J-5. Half and Aligned Sections. References: §§ 9.7, 9.13, 13.17. Draw required sections with instruments, using a sharp F lead for visible lines and a sharp 2H lead for sectioned lines. Include rounds and fillets, and, if assigned, add finish marks (∨ or ✗ as assigned) in all views.

J-6. Half and Aligned Sections. References: §§9.7, 9.11, 9.13, 13.17. Draw required sections, using instruments. If assigned, add finish marks (∨ or ✗ as assigned) in all views of Prob. 1.

J-7. Full and Half Sections. References: §§9.1, 9.7, 16.21. Use CI (cast iron) section lining in both problems. In Prob. 2, the right half of the assembly is to be left in elevation (i.e., not in section) and the left half is to be completed as a half section. Use a sharp F lead for visible lines and a sharp 2H lead for section lines. To draw the conventional break, see Fig. 9.37.

J-8. Ribs and Full Section. References: §§9.1, 9.10, 9.12, 13.17. Use a fairly sharp F lead for visible lines and a sharp 2H lead for section lines and center lines. Add finish marks in the sectional view of Prob. 1, and draw fillets and rounds neatly to match those in the given views.

J-9. Broken-Out and Removed Sections. References: §§9.8, 9.10, 9.11. Draw required sections. Use a fairly sharp F lead for visible lines and a sharp 2H lead for section lines and center lines. Draw break line in Prob. 1 as shown in Fig. 2.16; then draw the broken-out section. In the removed sections, show all visible lines behind the cutting plane in each case. Carefully draw the small fillets and rounds freehand, §7.34. Add finish marks in both problems.

J-10. Aligned Sections. References: §§9.13, 13.17, Fig. 7.38 (c). Use a fairly sharp F lead for visible lines a sharp 2H lead for section lines and center lines. Add finish marks in the sectional views.

GROUP K. AUXILIARY VIEWS

K-1. Auxiliary Views. References: §§6.5–6.7, 10.1–10.5. Sketch indicated auxiliary views, using a fairly sharp HB lead. In Prob. 1, number the points in the auxiliary views, making numbers .06" (1/16") high. In Probs. 2, 3 and 4, include all hidden lines. To transfer distances, make marks on the edge of a sheet of paper or a card.

K-2. Auxiliary Views. References: §§7.34–7.36, 10.1–10.11, 10.16, 13.17. Draw the required auxiliary views, showing the complete object in each problem. Use a fairly sharp F lead for visible lines and a sharp 2H lead for center lines. Draw fillets and rounds carefully freehand to match those given. In Prob. 1, the purpose is to show the true shape of the groove, while in Prob. 2, the objective is to show the true size and shape of an inclined surface. Show finish marks in all views in both problems, including the auxiliary views. Show reference-plane lines in both problems, §§10.4 and 10.5. The method of manipulating the triangle for drawing the parallel auxiliary-view projection lines and the perpendicular reference line in Prob. 1 is shown in Fig. 10.4.

In Prob. 1, in the view where the true angle between surfaces A and B is shown true size, indicate the actual number of degrees after measuring with a protractor. See Fig. 13.17. In Prob. 2, use guide lines, and fill in dimensions with letters .12" (1/8") high. Measure dimensions to nearest millimeter or equivalent decimal inch as assigned.

K-3. Auxiliary Views. References: §§10.1–10.11, 10.13–10.17. Sketch freehand or draw, if assigned, all missing lines in these problems. Lines may be missing in any of the views. Use a fairly sharp F lead for visible lines and hidden lines and a sharp 2H lead for center lines. Omit hidden lines in the auxiliary sections in Probs. 7 and 9.

K-4. Auxiliary Views. References: §§7.23, 7.24, 10.1–10.6, 10.11, 10.13. Draw the required auxiliary view and left-side view. Number, in all views, the corners of the *oblique surface*, using guide lines and making your numbers the same size as those given. Omit finish marks, as material is CRS (cold-rolled steel), and the object is understood to be finished all over.

K-5. Auxiliary Views. References: §§7.9, 10.1–10.7, 10.11. Draw the required auxiliary views, using instruments. Omit finish marks, since material is CRS and the objects are understood to be finished all over. In Prob. 1, letter the numbers in the auxiliary view, making the numbers the same size as those given. In both problems, indicate the required angles with arcs terminated by arrowheads, and the angles in degrees (use protractor), Fig.13.17.

K-6. Auxiliary Section. References: §§7.33–7.36, 10.1–10.10, 10.17. Draw required auxiliary section, using a fairly sharp F lead for visible lines and a sharp 2H lead for section lines and center lines. The shop notes are explained in §§7.33. Note that the edges of the cored hole are rounded. Observe that in the auxiliary view the object will be in an inverted position with the bottom of the object *up*. Show finish marks, §13.17. The smaller rounded and filleted edges should appear as conventional edges in the auxiliary view, §7.36.

K-7. Secondary Auxiliary View. References: §§10.7, 10.11, 10.14, 10.19–10.24. First, draw the partial primary auxiliary view as indicated by arrow A, using break line as in Fig. 10.15. Draw a complete depth auxiliary view of the entire object—about 19 mm (.75") from the front view—showing the true circular view of the tapped holes. Then draw a complete secondary auxiliary view of the entire object showing the true right section of the groove, as shown by arrow B in the pictorial. Show hidden lines in all views. The circular views of the tapped holes are drawn as shown in Fig. 15.8.

GROUP L. REVOLUTIONS

L-1. Revolutions. References: §§11.1, 11.2. Draw the required views. In Prob. 2, to obtain depth measurements for 20° and 30° angles in the top and side views, draw construction-line partial auxiliary view of the groove, projected upward to the left from the front view. This construction will fall on printed matter, but that is satisfactory.

L-2. Successive Revolutions. References: §§11.1–11.5. Draw required views as indicated, using a sharp F lead for visible lines and hidden lines. Make numbers to match those shown.

GROUP M. ISOMETRIC DRAWING

M-1. Isometric Drawing. References: §§18.3–18.14. In each of the two spaces draw an assigned problem, using a fairly sharp F lead for visible lines and a sharp 4H lead for construction lines. *Do not erase construction lines.* Omit hidden lines except in Prob. 1 (b). Start isometrics at given corner points. For all problems in Space 2, study the method of constructing angles in §18.14. Show construction triangles on your drawing.

M-2. Isometric Drawing. References: §§18.6–18.15, 18.18–18.20. Draw one assigned problem, starting at the given corner point. All small fillets and rounds are to be ignored and the edges drawn sharp.

M-3. Isometric-Irregular Combinations. References: §§18.18–18.20. Draw isometric of either assigned problem, taking dimensions from the given views with dividers, and doubling them on the isometric. Omit hidden lines unless necessary for clearness.

M-4. Isometric Sections. References: §§18.18–18.22, 18.24, 20.18. Draw assigned problem, starting the enclosing box at the given corner point. Omit hidden lines. In Prob. 6, use alternate section lining for rib, Fig. 9.27. In Probs. 4 and 6, show fillets and rounds, and shade them.

GROUP N. OBLIQUE PROJECTION

N-1. Oblique Projection. References: §§19.1–19.6, 19.9. Draw assigned problem in each space. Omit hidden lines unless necessary for clearness. Show all constructions clearly, and do not erase construction lines.

N-2. Oblique Projection. References: §§19.3–19.7. Draw assigned problem in each space. Omit hidden lines unless necessary for clearness. Show all construction lines clearly, and do not erase construction lines. In Prob. 2 (b), the ellipse construction must be drawn by means of offsets, §19.8. Show construction for all points of tangency.

N-3. Oblique Projection. References: §§19.3–19.10. Draw oblique views or sections as indicated. Show all construction lines. In Prob. 1, all small fillets and rounds are to be ignored and the edges drawn sharp. For Prob. 4, see Fig. 9.1 (b) as an example of an exploded oblique drawing. Draw the halves 25 mm (1.00") apart in Prob. 4.

GROUP O. THREADS AND FASTENERS

O-1. Detailed Unified Threads. References: §§15.1–15.7, 15.11, 15.12, 15.19, 15.20.
  Prob. 1. Draw the indicated threads, using detailed representation. The recess in the center of the longitudinal view is a Pratt & Whitney keyseat, which is cut into the threads. Note the tolerances on the major diameters of the threads. Be sure to finish the end view.
  Prob. 2. For the threads at the left end, the easiest method to space the crests is to use the scale directly. In the nut, the threads go all the way through. This portion should be drawn in a manner similar to Fig. 15.6 (b). Complete the section lining, using a sharp 2H lead. The thread pitches at the right end should be set off with the scale.
  Alternate Prob. 2. Same as Prob. 2 above, except draw square threads at left end as indicated. Study Fig. 15.13 to note the differences in the three parts of the threaded connection.

O-2. Thread Symbols. References: §§15.7–15.10, 15.19–15.21, 15.24. Draw schematic or simplified thread symbols as indicated or assigned.
  Prob. 1. Draw M8 tapped hole and M16 external thread in all three views as shown in Figs. 15.7 and 15.9. Add the tap drill size in the note, giving the diameter in millimeters. See Appendix 15. Be sure your bow-pencil lead is carefully sharpened and adjusted , and take care to make circular hidden-line dashes uniformly and carefully spaced. Note that since there is no relief, the specified thread length of 25 mm is not exact. Draw your last crest line at the 25 mm mark or slightly to the left of it. The 25 mm includes the chamfer. Make lines of the relative weights shown in Fig. 15.9. Draw conventional intersection of M8 tapped hole and small drilled hole, Fig. 7.38 (a).
  Prob. 2. First, chamfer the ends of the threaded screw, Fig. 15.9. Look up thread pitch in Appendix 15. See §15.21 for thread notes, which are to be lettered above the screw as indicated. Show threads in the end view. See Fig. 13.49 (a) for an example of knurling. The threads

at the right end of the screw will be sectioned similar to Fig. 15.7. Note that to determine the thread depth use the methods described in Fig. 15.9.
  Prob. 3. This is a portion of an assembly drawing. Study Figs. 16.86 and 16.93 for typical examples of schematic threads in assembly. Draw elevation view of eye bolt thread in the tapped hole. Show chamfer on end of eye bolt, and the relief at the top of the threaded portion when the screw is assembled in the tapped hole. Note that the scale is half size.
  Prob. 4. In the view, show the tapped hole in section with the screw in place similar to Fig. 15.17.

O-3. Square Threads. References: §§15.3–15.7, 15.12, 16.21. Draw square threads as indicated. Study Fig. 15.13, which shows how the threads mate in assembly. There are three distinct aspects of the thread tooth, or profile, as shown in the three parts of the figure. Complete all section lining, using a sharp 2H lead, but do not section the Jack Screw. Section the Jack Screw Nut up to the break line near the top.

O-4. Acme Threads. References: §§15.3–15.7, 15.13. First, identify the parts shown in the assembly. It will be seen that the Brake Shoes (8994) have cored holes to receive the two pieces (8995) of hard composition material. When the Handle (7025) on the Feed Nut Adjusting Screw (7023) is turned clockwise (looking down), the two Brake Shoes will move toward each other to act as a brake on the V-Belt Pulley (9033). Note that the scale for the details is double size, and that the threads must be symmetrical about the central collar of the screw. Start the two threads on the screw at corresponding points symmetrically on each side of the collar, and draw them away from the collar uniformly in opposite directions.

O-5. American Standard Bolts and Nuts. References: §§15.23–15.27. Draw the two bolts and nuts as indicated, using a sharp 4H lead for construction lines and a sharp F lead for visible lines. Draw detailed threads where the bolt diameter is over 1.00" or 25 mm. To draw the bolts, use the methods of Figs. 15.29 and 15.30. Show nuts screwed against right-hand plate. Complete the section lining, using a sharp 2H lead.

O-6. Fasteners. References: §§9.8, 15.28–15.34, 16.21.
  Prob. 1. See Fig. 15.34 (a) for the American National Standard Square Head Set Screw, but use the "full dog" point shown at (h). Sink this full dog point into a flat-bottomed 1/2" dia. × 1/4" deep hole in the Cap. See §15.28 for the information on the Jam Nut. This Jam Nut is chamfered on one side only, and has a washer face on the other. It is drawn like a regular hexagon finished nut, Fig. 15.29, except that it is thinner. Section the Body and the Spring and the broken-out portion of the Cap. See Fig. 15.44 (a). Use CI section lining for all parts.
  Prob. 2. For the 1/4" Hex. Socket Head Cap Screw, see Fig. 15.32 (e). In addition, give the screw a thread length of 5/8". It fits into a drilled and counterbored hole in the manner shown in the figure. For the Woodruff Key, see §15.34 and Appendix 23. For the Jam Nut, see

§15.28. Complete the assembly in section, observing the rules in §16.21. Use CI section lining for all parts. Do not section screws.

O-7. Fasteners. References: §§15.28–15.34, 16.21.

Prob. 1. For the 3/8" Cap Screw and Lock Washer, see §§15.28 and 15.29. See Fig. 15.31 (c) for lock washer applications, and Appendix 27 for dimensions. Select lock washer next size large than shank of screw. Complete the assembly in section, using CI section lining for all parts.

Prob. 2. For the 5/16" Headless Set Screw, see Fig. 15.34. For the 3/8" Fill. Head Cap Screw, and the 5/16" Flat Head Cap Screw, see §15.29. Complete the assembly in section, using CI section lining for all parts. Do not section screws.

GROUP P. SPRINGS

P-1. Springs. References: §§15.37, 15.38.

Prob. 1. A pictorial drawing of a round wire compression spring having 5 total coils is shown in Fig. 15.50 (a). The steps in drawing a detailed elevation view of this spring are shown at (d), (e), and (g). In Prob. 1, draw a similar spring, but with 6 total coils instead of 5. As indicated in the instructions, use Washburn & Moen Standard steel wire, and fill in the blank with the diameter of the wire (see Appendix 28). Show all hidden lines. Do not erase construction lines.

Prob. 2. The construction for a schematic elevation of a compression spring having 6 total coils is shown in Fig. 15.49 (a). In Prob. 2, draw a similar spring, but with 8 total coils instead of 6. Make lines the same weight as visible lines, using a fairly sharp F lead. Note in Fig. 15.49 (b) the construction which would be used if the total number of coils were a fractional number such as 6½.

Prob. 3. The construction for a schematic elevation view of an extension spring with 6 active coils and loop ends is shown in Fig. 15.49 (c). In Prob. 3, draw a similar spring, but with 7 active coils instead of 6. Make lines the same weight as visible lines, using a fairly sharp F lead.

Prob. 4. In Fig. 15.50 (b), The spring of (a) is shown with a cutting plane passing through the center line of the spring. At (c) The cutting plane is removed, showing the back half of the spring. The sectional view of the spring is shown at (f). In Prob. 4, draw a similar spring in section, but with square steel wire instead of round. Don't erase construction lines. See Fig. 15.44 (b).

Prob. 5. In Fig. 15.50 (h) is shown the construction for a detailed round wire compression spring having a fractional number of total coils (5½). In Prob. 5, draw a similar spring in section, but with 7½ total coils instead of 5½. Refer to Appendix 28, and fill in the diameter of the wire as indicated. Do not show hidden lines for the portion of the spring that falls behind the hub above point B in the drawing, but show all visible lines in the sectional view. Add missing visible lines in the assembly adjacent to the spring. Use steel section lining.

Note:

(1) Most helical springs are LH wound.

(2) Pitch of a spring is seldom given on industrial drawings. Instead, number of coils is given.

(3) Springs are closed and ground square to eliminate bending in compression.

GROUP Q. DIMENSIONING

Q-1. Dimensioning. References: §§4.20, 7.33, 13.1–13.31, 13.43. Dimension the drawings completely. Use the unidirectional decimal metric or decimal-inch system as assigned, Figs. 13.1, 13.9, 13.10, 13.15. Scale the views to obtain dimensions. Use a sharp 2H lead for dimension lines, extension lines, and center lines. Use a sharp F lead for arrowheads and lettering. Space dimensions 10 mm (.40") from object outline and 10 mm (.40") apart. Use complete guide lines (4H lead) for all dimension figures and notes, as shown in Fig. 4.27, including vertical or incline guide lines. Make whole numbers and capital letters 3 mm (.12") high, and fractions 6 mm (.25") high.

Prob. 1. Dimension to a full-size scale, §§2.25–2.27. For locating the rectangular stop and the drill hole, refer to Fig. 13.33. For the drill note, see Fig. 7.40 (a). Add note FAO (finish all over) at the bottom.

Prob. 2. Dimension to half-size scale, §§2.25–2.27. For the two holes, see Figs. 7.40 (d) and 13.44 (c). Add note **2 HOLES** at end of note. Locate the right-hand hole as shown in Fig. 13.33 (b), and then locate the other hole from it. Dimension the angles as shown in Fig. 13.17 (a).

Q-2. Dimensioning. References: §§4.20, 7.33, 13.1–13.31, 13.43. Dimension the drawings completely. Use the unidirectional metric or decimal-inch system, Figs. 13.1, 13.9, 13.10, 13.15. Use a sharp 2H lead for dimension lines, extension lines, and center lines. Use a sharp F lead for arrowheads and lettering. Space dimensions 10 mm (.40") from object outline and 10 mm (.40") apart. Use complete guide lines (4H lead) for all dimension figures and notes, as shown in Fig. 4.27, including vertical or inclined guide lines. Make whole numbers and capitals 3 mm (.12") high, and fractions 6 mm (.25") high.

Prob. 1. See Appendix 19 for the 5/16" hexagon socket cap screws. See Fig. 7.40 (c) for a typical drill and counterbore note. Identify *mating dimensions*, §13.26, by lettering a small M in a circle near the dimension figure, thus: Ⓜ Omit finish marks, as material is CRS and understood to be finished all over.

Prob. 2. Give drill note similar to upper note in Fig. 13.32 (a), but omit depth. Locate one small hole with respect to the center of the piece; then locate the second small hole from the first, §13.25. Give radii as shown in Fig. 13.19 (c). For the reamed hole, give note: **26 DRILL, 26.80–26.83 REAM.** This includes the specified 0.8 mm clearance.

Q-3. Dimensioning. References: §§4.20, 13.1–13.31, 13.43, 14.8. Dimension the drawings completely. Use the unidirectional metric system, Figs. 13.1, 13.10, 13.15. Space dimensions 10 mm from the object outline and 10 mm apart. Use complete guide lines for all dimension figures and notes. Make whole numbers and capitals 3 mm (.12") high.

Prob. 1. For the drill holes, give note similar to Fig. 13.36 (a). For the counterbored hole, give diameter at left of side view in millimeters to two places. For the reamed hole give note leading to circular view of the hole. For the RC 5 fit, see Appendix 5 and convert to metric measurements. For explanation of use of tables, see §14.8. In completing the right-side view, show only the extreme holes on the center lines near the edge of the flange. The other holes should be omitted, as their hidden lines would detract from the clearness of the drawing.

Prob.2. Dimension to half-size scale, §2.25. Dimension positive cylinders with respect to each other by a dimension between their center lines. The large hole is to be reamed, and a note to be given similar to Fig. 7.40 (b). For the RC 2 fit, see Appendix 5 and convert to metric measurements. For the keyway, refer to Appendix 21 for the exact size based on the diameter 29 mm (1 1/8"). The keyway is for a square key. Dimension the keyway as in Fig.13.48 (c), except give single millimeter values to two decimal places (without tolerances). Show your calculations in the margin at the top of the sheet. See Fig. 13.48 (d).

Q-4. Limit Dimensioning. References: §§13.22, 13.24, 13.31, 13.34, 13.43, 14.1–14.10, 15.21. Dimension the details completely. Use the unidirectional metric or decimal-inch system as assigned. Use two-place inch decimals or one-place millimeter decimals for all except the limit dimensions specified. Use complete guide lines, Fig. 4.27, making whole numbers and notes 3 mm (.12") high. For limit dimensions, see Fig. 13.12 (b)

The required American National Standard fits are indicated on the small assembly. For fit "A" give tolerances of 0.05 mm (.002") and an allowance of 0.08 mm (.003"), with the allowance subtracted from the hub and the bushing. The handwheel, shaft, and pinion should revolve freely where the nuts are securely tightened.

For Woodruff keyseats, give one note: **No. 505 WOODRUFF KEYSEATS.** See §15.34 and Appendix 23. For threads, give one note leading to either end of front view, but add: **BOTH ENDS.** Use equivalent metric threads or Unified Fine Series Threads. See Fig. 15.19 and Appendix 15. For chamfers give one note similar to Fig. 13.46 (b), but add: **BOTH ENDS.** For fillets, give one radius similar to Fig. 13.19 (e).

Q-5. Dimensioning. References: §§4.20, 7.33, 13.1–13.31, 13.34, 13.43, 14.1–14.8. Assume that this is a small production job of 100 units in which few drill jigs or fixtures will be used. Under or near each detail give the part number in a circle, followed by the part name. Under the part name give material and number required. Use complete guide lines for notes and dimensions, as shown in the upper left corner of the sheet. Space dimension lines 10 mm (.40") from the object outline and 10 mm (.40") apart. For mating parts, be sure to give corresponding mating dimensions, §13.26. Use unidirectional metric or decimal-inch dimensions as assigned.

Part 5. Hinge Pin. Brass—1 Req'd. Made from 6.5 mm (.25") dia. brass rod and cut to 45 mm (1.75") lengths by placing rods in bench vise and sawing with hand hack saw or band saw.

Part 6. Clamp Screw Pin. CRS—1 Req'd. Made from 3.5 mm (.13") dia. drill rod and cut to 19 mm (.75") lengths as for Part 5.

Part 4. Clamp Screw. Screwstock—1 Req'd. Made from 13 mm (.50") dia. free-cutting screw stock in hand screw machine.
Operations:
  (1) Feed stock through chuck to stop plug.
  (2) Turn down 10 mm (.40") dia. portion.
  (3) Chamfer 1.6 mm (.06") × 45°.
  (4) Thread: M10 × 1.5 or 3/8"–16 UNC–2A, 10 mm (.40") from end. (Drill 3.5 mm (.13") hole in assembly.)
  (5) Cut off to 60 mm (2.38") long.
  (6) Turn around and re-chuck in collet to stop plug.
  (7) Chamfer large end 1.6 mm (.06") × 45°.
  (8) Thread: M12 × 1.75 or 1/2"–13 UNC–2A.

Part 3. Clamp Screw Handle. CRS—1 Req'd. Made from 19 mm (.75") dia. CRS stock in hand screw machine.
Operations:
  (1) Feed stock through chuck 13 mm (.50") to stop plug.
  (2) Spot center with center drill.
  (3) 8.5 mm (.312") drill, 13 mm (.50") deep for tapped hole.
  (4) Plug tap M10 × 1.5 or 3/8"–16 UNC–2B, 10 mm (.40") deep.
  (5) Feed stock full length to stop plug.
  (6) Form chamfered shoulder, and turn shank to 10 mm (.40") dia. with swing tool.
  (7) Form ball with forming tool, leaving 6.5 mm (.25") dia. at end.
  (8) Cut off with cut-off tool to form profile at end of ball.
  (9) Drill 3.5 mm (.13") hole 5 mm (.20") from end.
 (10) Bend handle through 51° in fixture.

In Assembly: Assemble Parts 2, 3, and 4. Drill 3.5 mm (.13") hole in Part 4, using 3.5 mm (.13") hole in Part 3 as pilot.

Q-6. Dimensioning. References and preliminary instructions same as for Q-5. In addition, study §7.34. Show all finish marks. For bearing, use RC 6 fit, Appendix 5, and convert to metric measurements if assigned. For mating parts, be sure to give corresponding mating dimensions.

Part 2. Bearing Clamp Cap. CI—1 Req'd. At the bottom of sheet give title as for previous parts on Q-5. Assume that casting will be tumbled and cleaned and that pattern will be made of two pieces (one main piece and small lug on bottom).
Operations:
  (1) Straddle mill right end of bottom. For the width of the hinge: allowance 0.05 mm (.002") applied negatively) and tolerance 0.05 mm (.002").
  (2) Move table over and mill left end at same level.
  (3) 14.5 mm (.562) counterdrill large hole at left end, 22 mm (.88") deep.

(4) 10.3 mm (.41") small drill through.
(5) 3.2 mm (.12") drill oil hole.
In Assembly: Clamp Parts 1 and 2 together with 0.25 mm (.010") shim, and drill and ream bearing hole for RC 6 fit, Appendix 5. Drill and ream hinge pin hole for class RC 8 fit, Appendix 5. Convert Running Fits to metric measurements if assigned.

Q-7. Dimensioning. References and preliminary instructions same as for Q-5. In addition, study §7.34. Show all finish marks. For bearing, use RC 6 fit, Appendix 5, and convert to metric measurements if assigned. For mating parts, be sure to give corresponding mating dimensions.
Part 1. Bearing Clamp Base. CI–1 Req'd. Use right side of sheet as bottom of drawing. Give title information in upper right corner. Assume that casting will be tumbled and cleaned, and that pattern will be made of one piece.
Operations:
(1) Mill bottom surfaces.
(2) Mill top surfaces.
(3) Drill and tap M12 × 1.75 or 1/2"–13 UNC–2B, 3 holes.
(4) Mill hinge slot with Part 2, using fit described for Part 2.
In Assembly: Clamp Parts 1 and 2 together with 0.25 mm (.010") shim, drill and ream hinge pin hole for RC 8 fit, and bore and ream bearing hole for RC 6 fit, Appendix 5. Convert fits to metric measurements if assigned.

GROUP R. DETAIL DRAWINGS

Alternate Assignment (Sheets R-1 to R-9): Using a CAD system, produce a hard-copy drawing of the problem assigned. Dimension completely.

R-1. Detail Drawings. References: §§7.33–7.36, 12.1–12.10, 12.20, 13.43. Draw the required views as indicate in each problem. Dimension fully in unidirectional decimal-inch or metric system as assigned. Space dimensions 10 mm (.40") from the object outline and 10 mm (.40") apart. See Fig. 13.1, and follow the methods shown in the figure. Encircle each dimension on the printed problem as it is lettered on your drawing, to make sure that all dimensions have been used. *Observe the following:* Dimensions are not always (and cannot be) placed on the given views in the same places they should be on your drawing. Move any dimension if its given position is not the best available on the final drawing. Before handing in your drawing, check the rules of practice in §13.43, and make any corrections necessary. In Prob. 4, the fillets and rounds are 3 mm (.12") R. For Prob. 3, study §7.30 before giving the note naming the curve indicated at the right end of the piece.

R-2. Detail Drawings. References and instructions same as for R-1. In Prob. 2, draw the required left-side view adjacent to the front view. In Prob. 3, with reference to the surfaces marked LAP, See Appendix 2.

R-3. Detail Drawings. References and instructions same as for R-1. Plot enough points on all plotted curves to estab-

lish accurately the path of each curve. Use the irregular curve, §2.54, to draw perfectly smooth curves. Show all hidden lines. In Prob. 3, several holes will project obliquely and will appear as ellipses. Use the approximate ellipse of Fig. 5.56 or an ellipse template, Fig. 5.55.

R-4. Detail Drawings. References and instructions same as for R-1. In addition, review §§9.1–9.7, 9.11. SAE 1112 refers to a steel specified in the Society of Automotive Engineers Handbook.

R-5. Detail Drawings. References and instructions same as for R-1. For Prob. 2, study §§9.1 and 10.17. In this problem, surfaces B and C are finished.

R-6. Detail Drawings. References and instructions same as for R-1. For Prob. 1, study also §7.32. Note that chamfers are often omitted in the circular view when they detract from the clearness of the drawing.

R-7. Detail Drawings. References and instructions same as for R-1. In addition, study §§10.1–10.17. For Probs. 1 and 2, study the assembly drawing to determine how each part functions. In Prob. 3, omit hidden lines in auxiliary view.

R-8. Detail Drawings. References and instructions same as for R-1. For Prob. 3, study also §§7.38 and 9.1. Draw the required views in first-angle projection, laying them out with the metric scale and dimensioning the drawing in millimeters, §2.25.

R-9. Detail Drawings. References and instructions same as for R-1. In addition, for Prob. 1, study §10.7. In Prob. 2, draw the required right-side view adjacent to front full section.

R-10. Detail Drawings. References: Chapters 6–10, 13–15, §16.8. Draw or sketch the necessary views of the object assigned. Select appropriate scale and sheet size. Dimension completely using metric or decimal-inch dimensions as assigned.
Alternate Assignment: Using a CAD system, produce a hard-copy mutiview drawing of the problem assigned. Dimension completely.

R-11. Detail Drawings. References: Chapters 6–10, 13–15, §16.8. Draw or sketch the necessary views of the object assigned. Select appropriate scale and sheet size. Dimension completely using metric or decimal-inch dimensions as assigned.
Alternate Assignment: Using a CAD system, produce a hard-copy multiview drawing of the problem assigned. Dimension completely.

GROUP S. DESIGN LAYOUTS

S-1. Design Layout. References: §§13.43, 14.1–14.9, 16.1–16.17. The layout is given half size. Draw complete detail working drawings of assigned parts, taking sizes directly from the sheet with dividers. Make all measurements to the centers of the ink lines. For each detail, make a sketch

10

showing the views you consider necessary, together with unidirectional metric or decimal-inch dimensions as assigned. Then obtain your instructor's approval and his instructions as to the paper to use for the mechanical drawing.

S-2. Design Layout. References and instructions same as for S-1. The layout is given full size. For the RC 4 fit, see Appendix 5.

GROUP T. CHARTS AND GRAPHS

T-1. Pie and Bar Charts. References: §§28.19, 26.20, 28.15, 28.16. Use a sharp 4H lead for construction lines; a sharp F lead is recommended for required lines and lettering.
Prob.1. The percentage share of total U.S. auto industry sales in 1978 was as follows.

General Motors	47.6%
Ford	22.9
Chrysler	10.1
American Motors	1.5
Imports	17.7
Volkswagen of America	.2

Divide the pie chart to illustrate the above data. Label and show the value for each sector. Use .12" engineering lettering. Balance the large sector as one unit symmetrically about a vertical center line in the lower area of the circle.
Prob. 2. Construct a bar chart for the following data on publications in science of mechanisms.

Publications in Science of Mechanisms

Year	*No. of publications*
1910	450
1920	650
1930	850
1940	1350
1950	2000
1960	3000
1970	4600
1980	5800 (est.)

Use .50" wide vertical bars beginning at the given horizontal base line. Allow .25" spaces at the beginning and end and between the bars. Since the bar for 1980 will extend beyond the given rectangle, "break" the bar as shown in Fig. 28.31 (e). Place the percentage values above each bar in .09" numerals and shade the bars as in Fig. 26.31 (a). Label the value for 1980 as estimated (EST). Draw horizontal grid lines for each 500 but do not draw them across the bars or through the value figures.

T-2. Engineering Graphs. References: §§28.3–28.9. Use a sharp 4H lead for construction lines; a sharp F lead is recommended for required lines and lettering. Sketch the curves in lightly and smoothly through the plotted points before finalizing with the French curve, §2.54, and/or a triangle.

Prob. 1. Plot the following test data on a general purpose snap-action switch on the rectangular coordinate grid. Show data points with small circles as in §28.6. Plot the Resistance (*R*) on the ordinate.

Resistance load (R), amps at 125 volts AC	Expected life millions of cycles)
2.5	5.3
5	4.4
10	3.0
15	2.0
20	1.6
25	1.1
30	.8
35	.5

Prob.2. Plot these same test data on the semilogarithmic coordinate grid. Show data points with small circles as in §28.6.

If assigned, find an empirical equation by the selected points method of §30.6 for the data as plotted in Prob. 2.

GROUP U. GRAPHICAL MATHEMATICS

U-1. Nomographs. References: §§29.8, 29.14, 29.15. Use a sharp 4H lead for construction lines; a sharp F lead is recommended for required lines and lettering.
Prob. 1. Complete the natural parallel-scale nomograph or alignment chart for the given equation. Space the scales 3.00" apart and make all scales same length as the given X-scale. Match height of scale numerals to given numerals and identify each scale. Check the diagram by substituting the following values: $x = 0$, $y = 8$, $x = 12$, $y = 12$ and the appropriate Z-scale values in the equation. Show calculations on the drawing. Start Z-scale at 0. Maximum Y value is 16.
Prob. 2. Complete a natural scale N-chart for the Ohm's law equation, $V = IR$, where *V* is voltage, *I* is current (scale range 0 to 10 amperes), and *R* is resistance (scale range 0 to 100 ohms). Make vertical scales of equal length and 6.00" apart. Match scale numeral heights to given numerals and label all scales. Check the nomograph by substituting the following values: 8 amperes, 50 ohms; 2.5 amperes, 80 ohms and the appropriate voltage values in the equation. Show these calculations on the drawing.

U-2. Empirical Equation. References: §§28.10, 30.1–30.7. Use a sharp 4H lead for construction lines; an F lead is recommended for the plotted curves and lettering. Make four plots of the given data:

(1) Use the linear scales for both variables.
(2) Use the linear scale for *x* and the logarithmic scale for *y*.
(3) Use the logarithmic scale for *x* and the linear scale for *y*.
(4) Use the logarithmic scales for *x* and *y*.

Distinguish each curve by using point symbols given in Fig. 28.14.

One of these plots should be most nearly a straight line. From it determine the appropriate empirical equation that relates the variables.

x	y
1.12	2
1.42	2.6
1.9	3.6
2.5	4.8
3.5	7
5.4	11
8	17

U-3. Simultaneous Equations. References: §§31.1–31.3. Use a sharp 4H lead for construction; an F lead is recommended for the curves and lettering. Plot the two equations and obtain their simultaneous solutions to three significant figures. It will be necessary to establish the coordinates of closely spaced points near the solution in order to obtain the desired accuracy. Substitute the values in the equations and compute the results to illustrate the accuracy of your solutions. Show the calculations on the drawing.

Determine the X-roots of the first equation to three significant figures. Label the solutions, the roots, and the curves.

U-4. Integral Calculus. References: §§31.12, 31.13. Plot the given data to describe the relationship of volume and pressure. Use a sharp 4H lead for construction, an F lead is recommended for the curves and lettering. Match the given lettering. Using the area law, integrate to determine the area under the pressure curve that is equivalent to the integral curve or in this case the work curve. The unit for work is in.-lb.

Pressure (p), lb/in.^{2} (psi)	Volume (V), in.^{3}
100	0
57	10
41	20
33	30
27.5	40
24	50
22	60

U-5. Differential Calculus. References: §§31.3–31.6. The given curve represents work versus time data. Using the slope law, plot the first derivative on the layout at the bottom of the sheet. Use a sharp 4H lead for construction; an F lead is recommended for the curves and lettering. Match the given lines and lettering. Since the work is measured in ft-lb per minute, the derived curve will show the power at any distance. See Table 31.1. Add the scale designations and caption for the ordinate axis of the derived curve. Fill in the blank for the time when the output is 1 horsepower.

GROUP V. COMPUTER-AIDED DRAFTING

V-1. Terms and Descriptions. References: Chapters 3 and 8, Appendix 3. Some terms related to computer graphics are given in the table. A list of descriptions for these terms is given on the right. Find the matching description for each term and enter its letter identifier in the table.

V-2. Two-Dimensional Coordinate Plot. Reference: Chapter 8.

Space 1. Digitize the single view drawing by defining the X and Y coordinates of the indicated points and fill in the given table. Point A is the origin with values of X and Y equal to zero. Consider each division of the grid as 1 unit. Keep in mind that any X values to the left of the origin and any Y values below the origin are negative.

Space 2. From the X and Y coordinate data given in the table, plot all points on the grid and complete the drawing. Point A is the origin. Consider each division of the grid as 1 unit.

V-3. Three-Dimensional Coordinate Plot. Reference: Chapter 8. In drawing an image, the actions of the pen are Move and Draw.

Move: The pen moves from its present position to new X, Y, and Z coordinates specified. A line is not drawn. Numeral 0 is used to indicate Move action.

Draw: A line is drawn from the present pen position to new X, Y, and Z coordinates specified. Numeral 1 is used to indicate Draw action.

Space 1. Determine X, Y, and Z coordinates for all the points of the object. Complete the table for drawing the object, starting with point A. Coordinates X, Y, and Z are positioned as indicated with arrows, with point A as origin. Try to use a minimum number of Move actions.

Space 2. According to the data shown in the table, draw the object on the grids provided. Coordinates X, Y, and Z are positioned as indicated by the arrows, with point A as origin.

V-4. Menu Usage. Reference: Chapter 8. The drawing shows the front view of a Bracket that is to be generated on a graphics terminal. The numbers 1 to 21 refer to graphic entities that make up the drawing. Available menu commands for generating entities are given on the right. Complete the table by determining the menu commands to generate the entities. Enter the letter identifiers (A, B, etc.) of menu selections in the table.

V-5. Coordinate Systems. Reference: Chapter 8. Using the given descriptions for VIEW COORDINATES and WORLD COORDINATES, complete the tables for the front and right-side views of the object. Point number 1 is considered as the origin. Each grid division is equal to 1 unit.

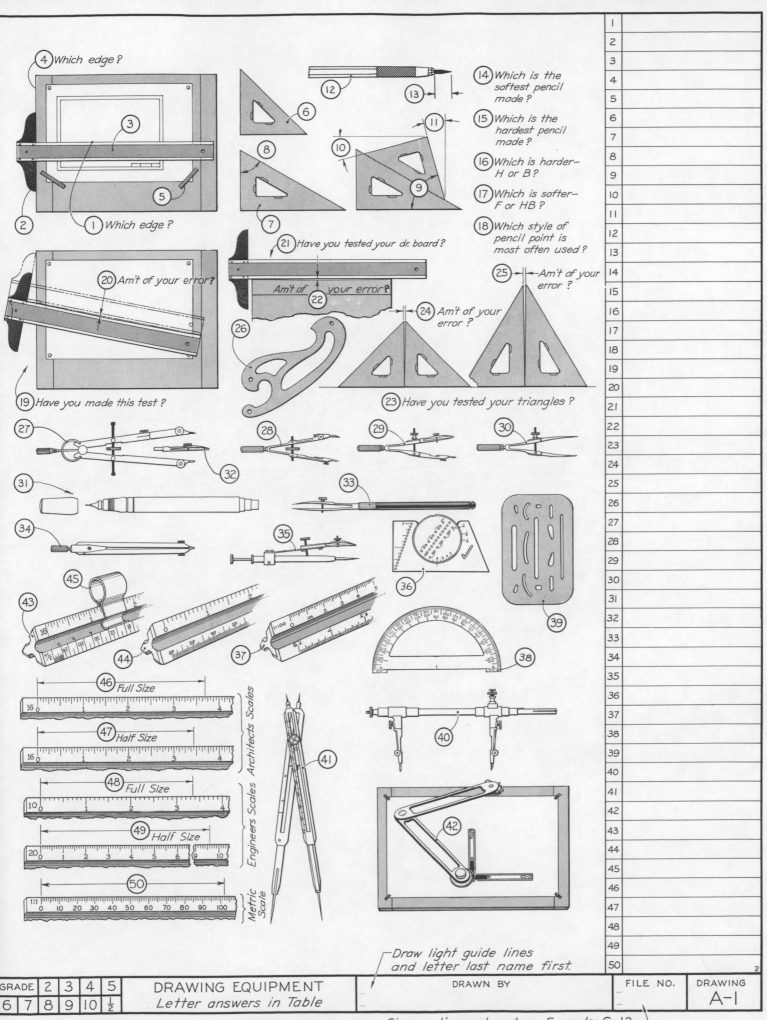

GRADE	2	3	4	5
	6	7	8	9	10	½

DRAWING EQUIPMENT
Letter answers in Table

DRAWN BY

FILE NO.

DRAWING
A-1

Give section and seat no. Example: C-12

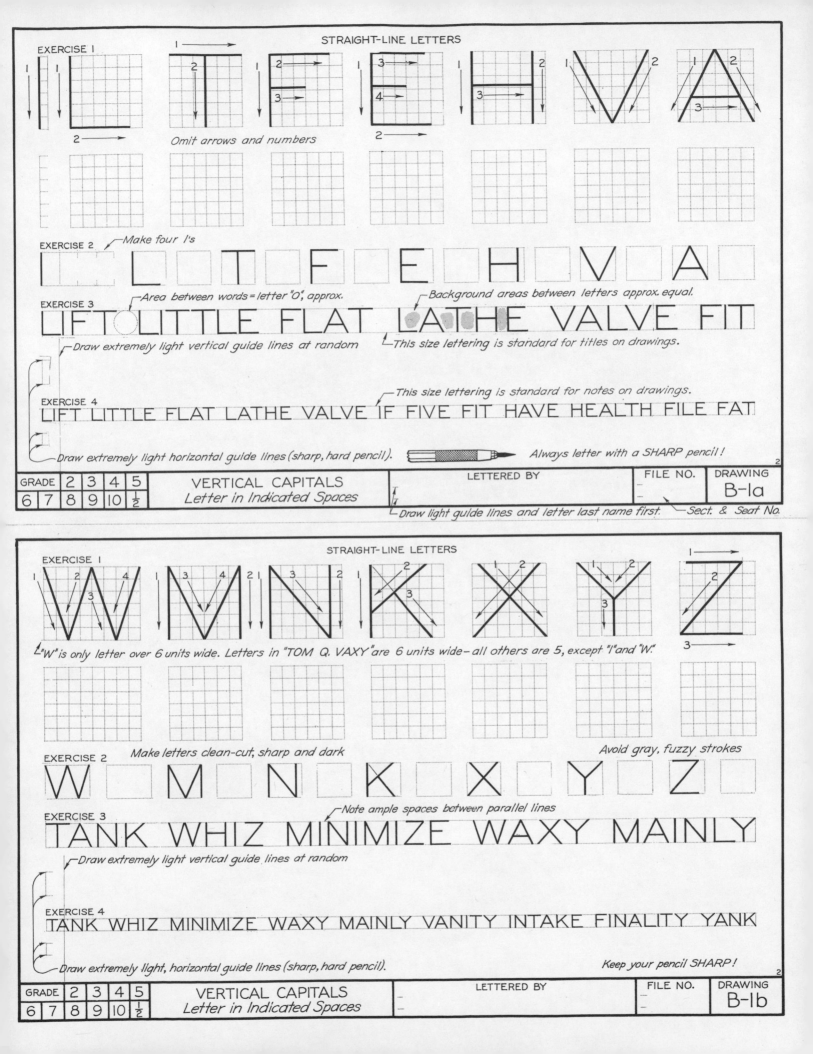

STRAIGHT-LINE LETTERS

EXERCISE 1

Omit arrows and numbers

EXERCISE 2 — Make four I's

I L T F E H V A

EXERCISE 3 — Area between words = letter "O", approx. — Background areas between letters approx. equal.

LIFT LITTLE FLAT LATHE VALVE FIT

— Draw extremely light vertical guide lines at random — This size lettering is standard for titles on drawings.

EXERCISE 4 — This size lettering is standard for notes on drawings.

LIFT LITTLE FLAT LATHE VALVE IF FIVE FIT HAVE HEALTH FILE FAT

— Draw extremely light horizontal guide lines (sharp, hard pencil). — Always letter with a SHARP pencil!

GRADE	2	3	4	5	VERTICAL CAPITALS	LETTERED BY	FILE NO.	DRAWING
6	7	8	9	10	½	Letter in Indicated Spaces			B-1a

— Draw light guide lines and letter last name first. — Sect. & Seat No.

STRAIGHT-LINE LETTERS

EXERCISE 1

W M N K X Y Z

— "W" is only letter over 6 units wide. Letters in "TOM Q. VAXY" are 6 units wide – all others are 5, except "I" and "W."

EXERCISE 2 — Make letters clean-cut, sharp and dark — Avoid gray, fuzzy strokes

W M N K X Y Z

EXERCISE 3 — Note ample spaces between parallel lines

TANK WHIZ MINIMIZE WAXY MAINLY

— Draw extremely light vertical guide lines at random

EXERCISE 4

TANK WHIZ MINIMIZE WAXY MAINLY VANITY INTAKE FINALITY YANK

— Draw extremely light, horizontal guide lines (sharp, hard pencil). — Keep your pencil SHARP!

GRADE	2	3	4	5	VERTICAL CAPITALS	LETTERED BY	FILE NO.	DRAWING
6	7	8	9	10	½	Letter in Indicated Spaces			B-1b

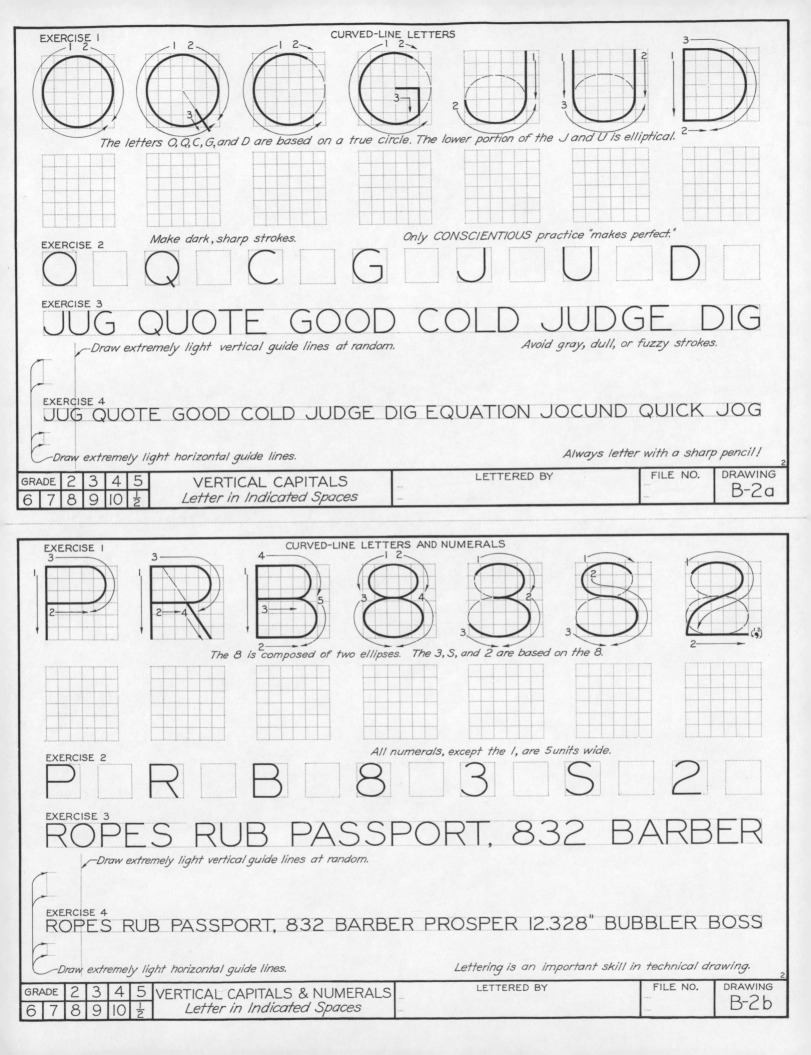

EXERCISE 1 — CURVED-LINE LETTERS

O O C G J U D

The letters O, Q, C, G, and D are based on a true circle. The lower portion of the J and U is elliptical.

Make dark, sharp strokes.   Only CONSCIENTIOUS practice "makes perfect."

EXERCISE 2

O   Q   C   G   J   U   D

EXERCISE 3

JUG QUOTE GOOD COLD JUDGE DIG

Draw extremely light vertical guide lines at random.   Avoid gray, dull, or fuzzy strokes.

EXERCISE 4

JUG QUOTE GOOD COLD JUDGE DIG EQUATION JOCUND QUICK JOG

Draw extremely light horizontal guide lines.   Always letter with a sharp pencil!

GRADE	2	3	4	5	VERTICAL CAPITALS	LETTERED BY	FILE NO.	DRAWING
	6	7	8	9	10	½	*Letter in Indicated Spaces*			B-2a

EXERCISE 1 — CURVED-LINE LETTERS AND NUMERALS

P R B 8 3 S 2

The 8 is composed of two ellipses.   The 3, S, and 2 are based on the 8.

All numerals, except the 1, are 5 units wide.

EXERCISE 2

P   R   B   8   3   S   2

EXERCISE 3

ROPES RUB PASSPORT, 832 BARBER

Draw extremely light vertical guide lines at random.

EXERCISE 4

ROPES RUB PASSPORT, 832 BARBER PROSPER 12.328" BUBBLER BOSS

Draw extremely light horizontal guide lines.   Lettering is an important skill in technical drawing.

GRADE	2	3	4	5	VERTICAL CAPITALS & NUMERALS	LETTERED BY	FILE NO.	DRAWING
	6	7	8	9	10	½	*Letter in Indicated Spaces*			B-2b

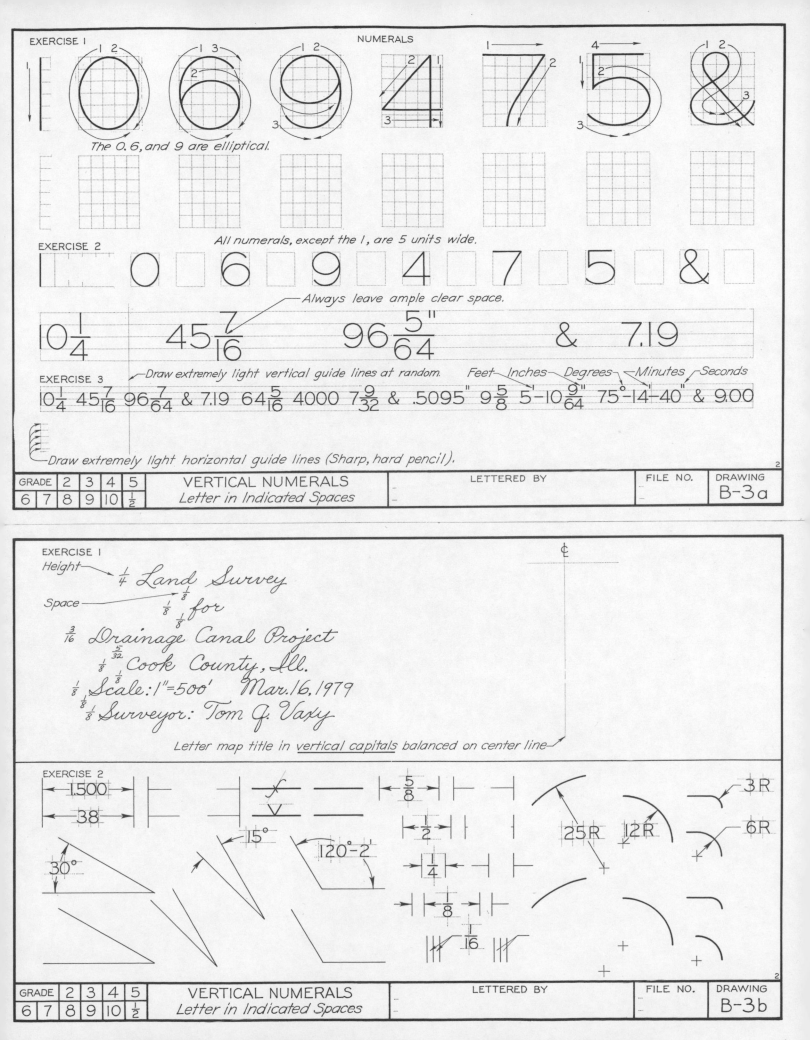

EXERCISE I   NUMERALS

The 0, 6, and 9 are elliptical.

EXERCISE 2

All numerals, except the 1, are 5 units wide.

$10\frac{1}{4}$ $45\frac{7}{16}$ $96\frac{5}{64}''$ $\&$ 7.19

Always leave ample clear space.

EXERCISE 3   Draw extremely light vertical guide lines at random.   Feet   Inches   Degrees   Minutes   Seconds

$10\frac{1}{4}$ $45\frac{7}{16}$ $96\frac{7}{64}$ $\&$ 7.19 $64\frac{5}{16}$ 4000 $7\frac{9}{32}$ $\&$ $.5095''$ $9\frac{5}{8}$ $5'-10\frac{9}{64}''$ $75°-14-40''$ $\&$ 9.00

Draw extremely light horizontal guide lines (Sharp, hard pencil).

GRADE	2	3	4	5	VERTICAL NUMERALS	LETTERED BY	FILE NO.	DRAWING
6	7	8	9	10	$\frac{1}{2}$	Letter in Indicated Spaces			B-3a

EXERCISE I

Height — $\frac{1}{4}$ Land Survey

Space — $\frac{1}{8}$
$\frac{1}{8}$ for

$\frac{3}{16}$ Drainage Canal Project
$\frac{5}{32}$
$\frac{1}{8}$ Cook County, Ill.
$\frac{1}{8}$
$\frac{1}{8}$ Scale: 1"=500'   Mar. 16, 1979
$\frac{1}{8}$
$\frac{1}{8}$ Surveyor: Tom G. Vaxy

Letter map title in <u>vertical capitals</u> balanced on center line

EXERCISE 2

1.500
38
15°
30°
120°-2
$\frac{5}{8}$
$\frac{1}{2}$
$\frac{1}{4}$
$\frac{1}{8}$
$\frac{1}{16}$
25R   12R
3R
6R

GRADE	2	3	4	5	VERTICAL NUMERALS	LETTERED BY	FILE NO.	DRAWING
6	7	8	9	10	$\frac{1}{2}$	Letter in Indicated Spaces			B-3b

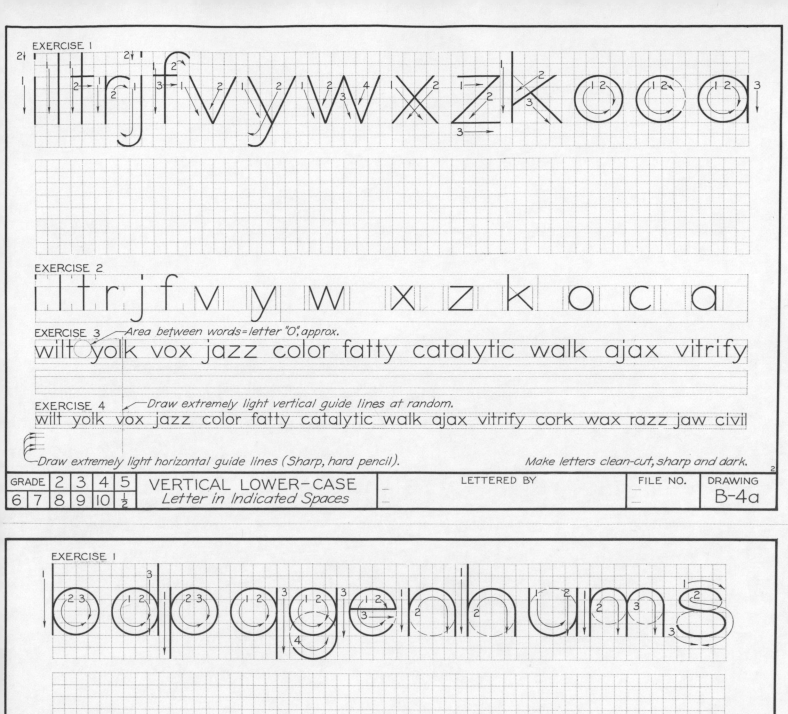

i l t r j f v y w x z k o c a

EXERCISE 2

i l t r j f v y w x z k o c a

EXERCISE 3 — *Area between words = letter "O", approx.*

wilt yolk vox jazz color fatty catalytic walk ajax vitrify

EXERCISE 4 — *Draw extremely light vertical guide lines at random.*

wilt yolk vox jazz color fatty catalytic walk ajax vitrify cork wax razz jaw civil

—Draw extremely light horizontal guide lines (Sharp, hard pencil).   *Make letters clean-cut, sharp and dark.*

GRADE	2	3	4	5	VERTICAL LOWER–CASE	LETTERED BY	FILE NO.	DRAWING
	6	7	8	9	10	½	*Letter in Indicated Spaces*	—	—	B-4a

EXERCISE 1

b d p q g e n h u m s

EXERCISE 2

b d p q g e n h u m s

EXERCISE 3

bump dens squeegee hums queue hung bud pump

EXERCISE 4 — *Draw extremely light vertical guide lines at random.*

whenever you tackle jobs requiring amazing dexterity, feel pride in your skills

—Draw extremely light horizontal guide lines.   *Accent ends of strokes.*

GRADE	2	3	4	5	VERTICAL LOWER–CASE	LETTERED BY	FILE NO.	DRAWING
	6	7	8	9	10	½	*Letter in Indicated Spaces*			B-4b

EXERCISE 1

ILTFEHVA

Omit arrows and numbers

EXERCISE 2 — Make four I's

I L T F E H V A

EXERCISE 3 — Area between words = letter "O," approx. — Background areas between letters approx. equal.

LIFT LITTLE FLAT LATHE VALVE FIT

— Draw extremely light inclined guide lines at random. — This size lettering is standard for titles on drawings.

EXERCISE 4 — This size lettering is standard for notes on drawings.

LIFT LITTLE FLAT LATHE VALVE IF FIVE FIT HAVE HEALTH FILE FAT

— Draw extremely light horizontal guide lines (sharp, hard pencil). — Always letter with a SHARP pencil!

GRADE	2	3	4	5	INCLINED CAPITALS	LETTERED BY	FILE NO.	DRAWING
	6	7	8	9	10	½	Letter in Indicated Spaces			B-5a

Draw light guide lines and letter last name first. Sect. & Seat No.

EXERCISE 1

WMNKXYZ

— "W" is only letter over 6 units wide. Letters in "TOM Q. VAXY" are 6 units wide – all others are 5, except "I" and "W".

EXERCISE 2 Make letters clean-cut, sharp, and dark. Avoid gray, fuzzy strokes.

W M N K X Y Z

EXERCISE 3 — Note ample space between parallel lines.

TANK WHIZ MINIMIZE WAXY MAINLY

— Draw extremely light inclined lines at random.

EXERCISE 4

TANK WHIZ MINIMIZE WAXY MAINLY VANITY INTAKE FINALITY YANK

— Draw extremely light, horizontal guide lines (Sharp, hard pencil). Keep your pencil SHARP!

GRADE	2	3	4	5	INCLINED CAPITALS	LETTERED BY	FILE NO.	DRAWING
	6	7	8	9	10	½	Letter in Indicated Spaces			B-5b

EXERCISE I

CURVED-LINE LETTERS

O O C G J U D

45°

The letters O, Q, C, G, and D are based on a true ellipse.   The lower portion of the J and U is elliptical.

EXERCISE 2

Make dark, sharp strokes.   Only CONSCIENTIOUS practice "makes perfect."

O   Q   C   G   J   U   D

EXERCISE 3

JUG QUOTE GOOD COLD JUDGE DIG

Draw extremely light inclined guide lines at random.   Avoid gray, dull, or fuzzy strokes.

EXERCISE 4

JUG QUOTE GOOD COLD JUDGE DIG EQUATION JOCUND QUICK JOG

Draw extremely light horizontal guide lines.   Always letter with a sharp pencil!

GRADE	2	3	4	5	INCLINED CAPITALS	LETTERED BY	FILE NO.	DRAWING
	6	7	8	9	10	½	Letter in Indicated Spaces			B-6a

EXERCISE I

CURVED-LINE LETTERS & NUMERALS

P R B 8 3 S 2

The 8 is composed of two ellipses. The 3, S, and 2 are based on the 8.

EXERCISE 2

All numerals, except the I, are 5 units wide.

P   R   B   8   3   S   2

EXERCISE 3

ROPES RUB PASSPORT 832 BARBER

Draw extremely light inclined guide lines at random.

EXERCISE 4

ROPES RUB PASSPORT, 832 BARBER PROSPER 12.328" BUBBLER BOSS

Draw extremely light horizontal guide lines.   Lettering is an important skill in technical drawing.

GRADE	2	3	4	5	INCLINED CAPITALS & NUMERALS	LETTERED BY	FILE NO.	DRAWING
	6	7	8	9	10	½	Letter in Indicated Spaces			B-6b

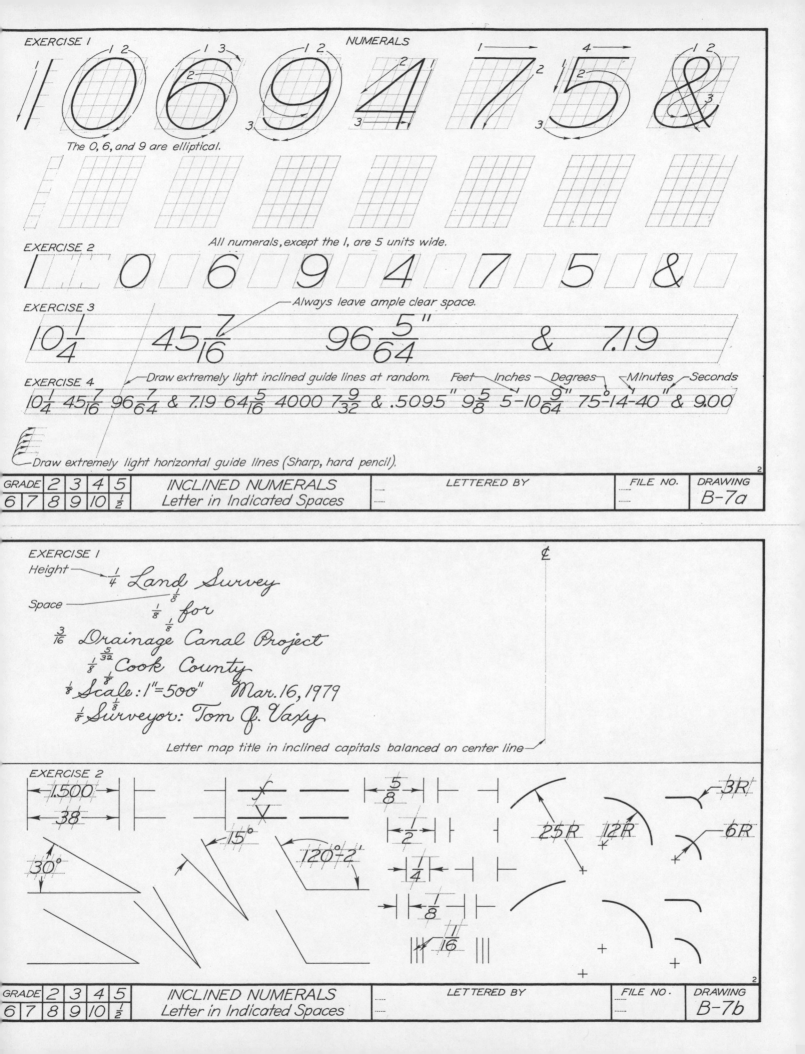

EXERCISE 1

NUMERALS

1 0 6 9 4 7 5 &

The 0, 6, and 9 are elliptical.

EXERCISE 2

All numerals, except the 1, are 5 units wide.

1 0 6 9 4 7 5 &

EXERCISE 3

Always leave ample clear space.

$10\frac{1}{4}$ $45\frac{7}{16}$ $96\frac{5}{64}$" & 7.19

EXERCISE 4

Draw extremely light inclined guide lines at random.

Feet Inches Degrees Minutes Seconds

$10\frac{1}{4}$ $45\frac{7}{16}$ $96\frac{7}{64}$ & 7.19 $64\frac{5}{16}$ 4000 $7\frac{9}{32}$ & .5095" $9\frac{5}{8}$ 5'-$10\frac{9}{64}$" 75°-14'-40" & 9.00

Draw extremely light horizontal guide lines (Sharp, hard pencil).

GRADE	2	3	4	5	INCLINED NUMERALS	LETTERED BY	FILE NO.	DRAWING
6	7	8	9	10	½ Letter in Indicated Spaces			B-7a

EXERCISE 1

Height — $\frac{1}{4}$ *Land Survey*

Space — $\frac{1}{8}$

$\frac{1}{8}$ *for*

$\frac{1}{8}$

$\frac{3}{16}$ *Drainage Canal Project*

$\frac{5}{32}$

$\frac{1}{8}$ *Cook County*

$\frac{1}{8}$

$\frac{1}{8}$ *Scale: 1"=500" Mar. 16, 1979*

$\frac{1}{8}$

$\frac{1}{8}$ *Surveyor: Tom J. Vaxy*

℄

Letter map title in inclined capitals balanced on center line

EXERCISE 2

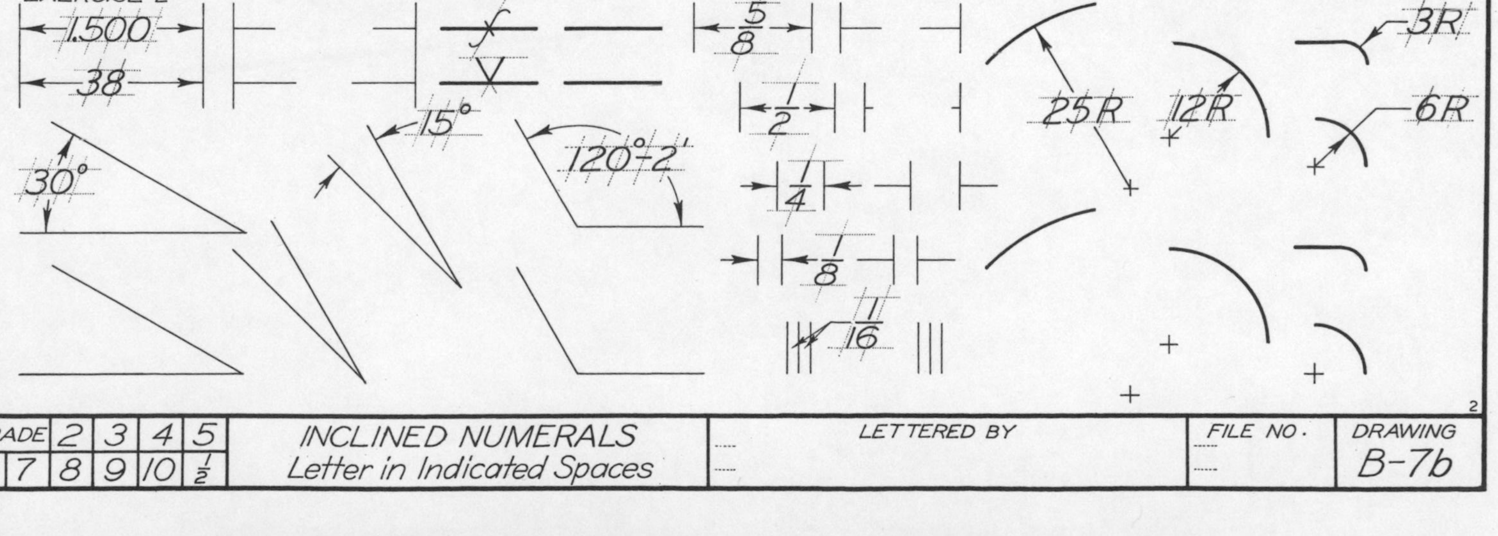

1.500
38
5/8
1/2
1/4
1/8
1/16
30°
15°
120°-2'
25R
12R
3R
6R

GRADE	2	3	4	5	INCLINED NUMERALS	LETTERED BY	FILE NO.	DRAWING
6	7	8	9	10	½ Letter in Indicated Spaces			B-7b

EXERCISE 1

i i l t r j f v y w x z k o o a o o a

EXERCISE 2

i l t r j f v y w x z k o c a

EXERCISE 3 — Area between words = letter "0", approx.

wilt yolk vox jazz color fatty catalytic walk ajax vitrify

EXERCISE 4 — Draw extremely light inclined guide lines at random.

wilt yolk vox jazz color fatty catalytic walk ajax vitrify cork wax razz jaw civil

Draw extremely light horizontal guide lines (Sharp, hard pencil).   Make letters clean-cut, sharp and dark.

GRADE	2	3	4	5	INCLINED LOWER-CASE	LETTERED BY	FILE NO.	DRAWING
	6	7	8	9	10	½	Letter in Indicated Spaces			B-8a

EXERCISE 1

b b d p q g e n h u m s

EXERCISE 2

b d p q g e n h u m s

EXERCISE 3

bump dens squeegee hums queue hung bud pumps

EXERCISE 4 — Draw extremely light inclined guide lines at random.

whenever you tackle a job requiring amazing dexterity, feel pride in your skills

Draw extremely light horizontal guide lines.   Accent ends of strokes.

GRADE	2	3	4	5	INCLINED LOWER-CASE	LETTERED BY	FILE NO.	DRAWING
	6	7	8	9	10	½	Letter in Indicated Spaces			B-8b

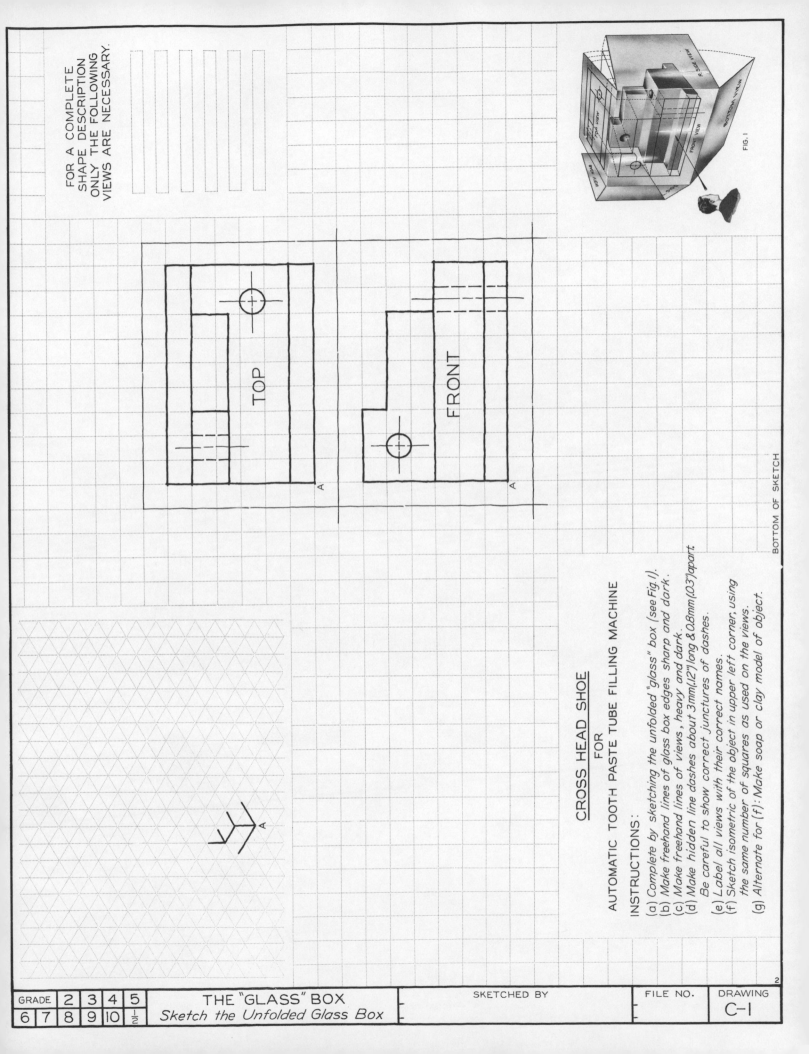

FIG. I

TOP

FRONT

A

A

A

BOTTOM OF SKETCH

CROSS HEAD SHOE

FOR

AUTOMATIC TOOTH PASTE TUBE FILLING MACHINE

INSTRUCTIONS:

(a) Complete by sketching the unfolded "glass" box (see Fig. I).
(b) Make freehand lines of glass box edges sharp and dark.
(c) Make freehand lines of views, heavy and dark.
(d) Make hidden line dashes about 3mm (.12") long & 0.8mm (.03") apart.
    Be careful to show correct junctures of dashes.
(e) Label all views with their correct names.
(f) Sketch isometric of the object in upper left corner, using
    the same number of squares as used on the views.
(g) Alternate for (f): Make soap or clay model of object.

GRADE	2	3	4	5
6	7	8	9	10	½

THE "GLASS" BOX
Sketch the Unfolded Glass Box

SKETCHED BY

FILE NO.

DRAWING
C-1

2

ROLLER REST
BRACKET
FOR
AUTOMATIC SCREW MACHINE

*Pictorial is not to any standard scale;
sketch views proportionately, by eye.
Include centerlines in sketch.*

BOTTOM OF SKETCH

Sketch front view here

GRADE	2	3	4	5	FREEHAND SKETCHING	SKETCHED BY	FILE NO.	DRAWING
6	7	8	9	10	2⁻	*Sketch Front, Top & Right Side Views*			C-4

2

1.

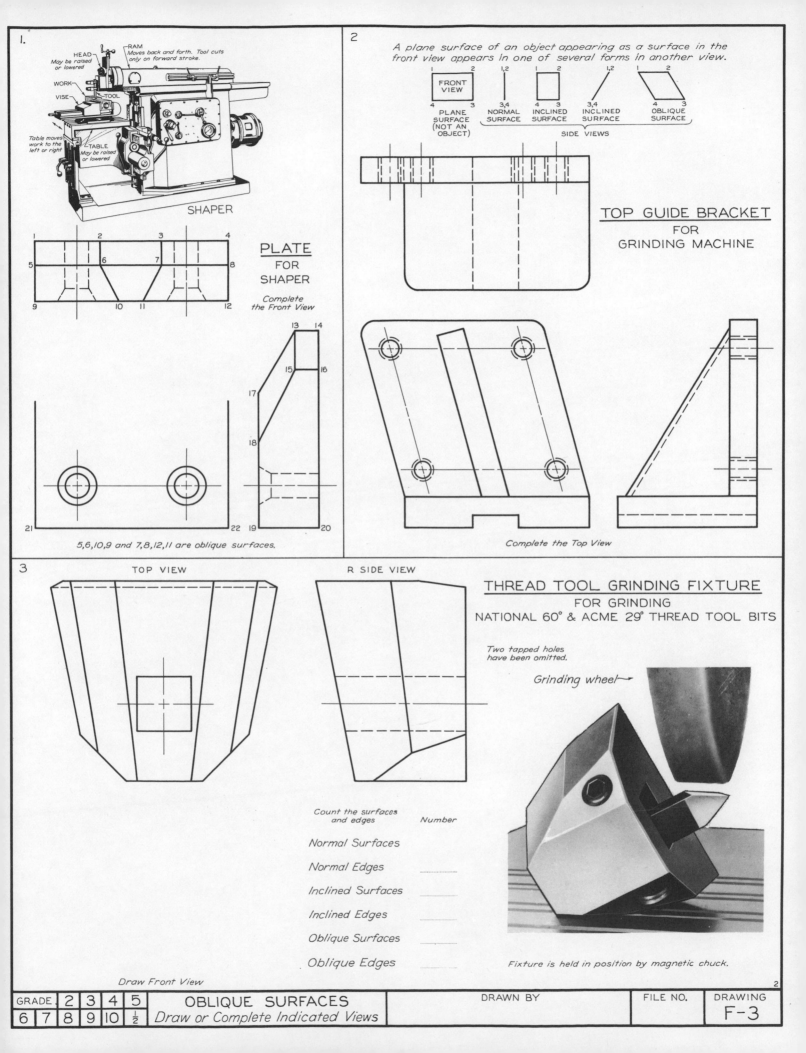

HEAD
May be raised or lowered

RAM
Moves back and forth. Tool cuts only on forward stroke.

WORK

VISE

TOOL

Table moves work to the left or right

TABLE
May be raised or lowered

SHAPER

PLATE
FOR
SHAPER

Complete the Front View

5,6,10,9 and 7,8,12,11 are oblique surfaces.

2

A plane surface of an object appearing as a surface in the front view appears in one of several forms in another view.

| FRONT VIEW | NORMAL SURFACE | INCLINED SURFACE | INCLINED SURFACE | OBLIQUE SURFACE |

PLANE SURFACE (NOT AN OBJECT)

SIDE VIEWS

TOP GUIDE BRACKET
FOR
GRINDING MACHINE

Complete the Top View

3

TOP VIEW

R SIDE VIEW

THREAD TOOL GRINDING FIXTURE
FOR GRINDING
NATIONAL 60° & ACME 29° THREAD TOOL BITS

Two tapped holes have been omitted.

Grinding wheel →

Count the surfaces and edges	Number
Normal Surfaces	
Normal Edges	
Inclined Surfaces	
Inclined Edges	
Oblique Surfaces	
Oblique Edges	

Fixture is held in position by magnetic chuck.

Draw Front View

GRADE	2	3	4	5
	6	7	8	9	10	½

OBLIQUE SURFACES
Draw or Complete Indicated Views

DRAWN BY

FILE NO.

DRAWING
F-3

I.

	FRONT	RIGHT SIDE
SURFACE	1-5	
EDGE	3-5	
SURFACE		24-26
SURFACE	8-9-15-14	
SURFACE		24-19
SURFACE	5-17	

Letter answers in the above table.

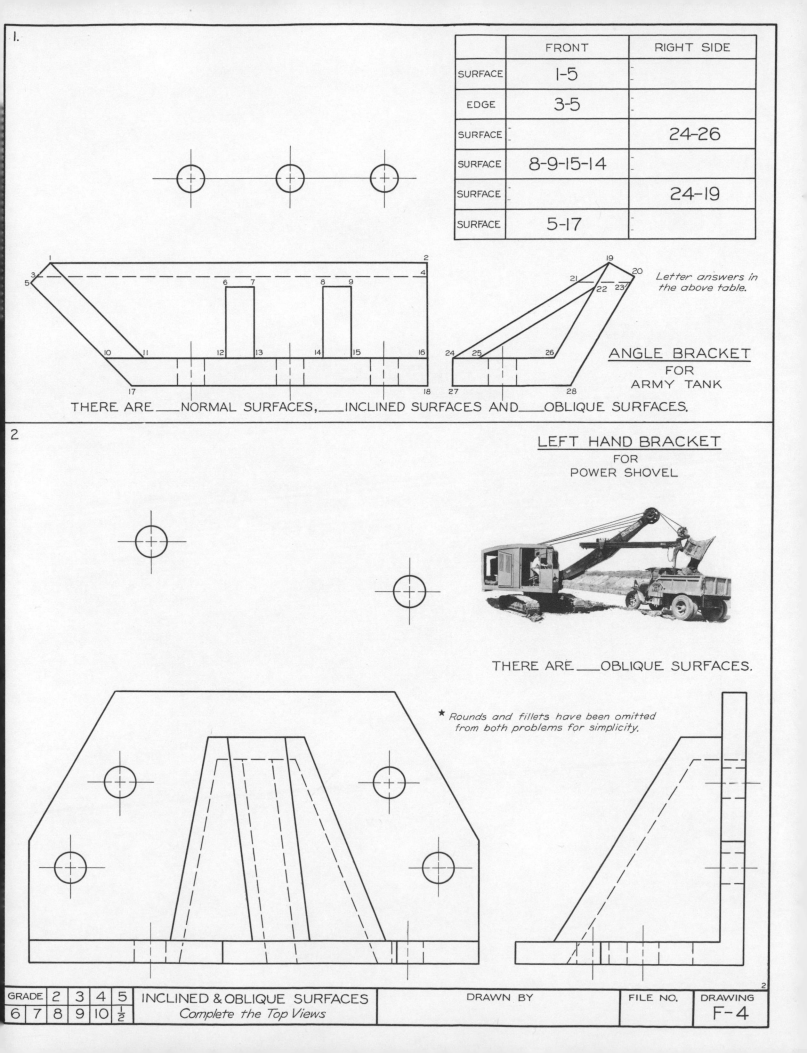

ANGLE BRACKET
FOR
ARMY TANK

THERE ARE ___ NORMAL SURFACES, ___ INCLINED SURFACES AND ___ OBLIQUE SURFACES.

2

LEFT HAND BRACKET
FOR
POWER SHOVEL

THERE ARE ___ OBLIQUE SURFACES.

★ *Rounds and fillets have been omitted from both problems for simplicity.*

GRADE	2	3	4	5	INCLINED & OBLIQUE SURFACES	DRAWN BY	FILE NO.	DRAWING
	6	7	8	9	10	½	*Complete the Top Views*			F-4

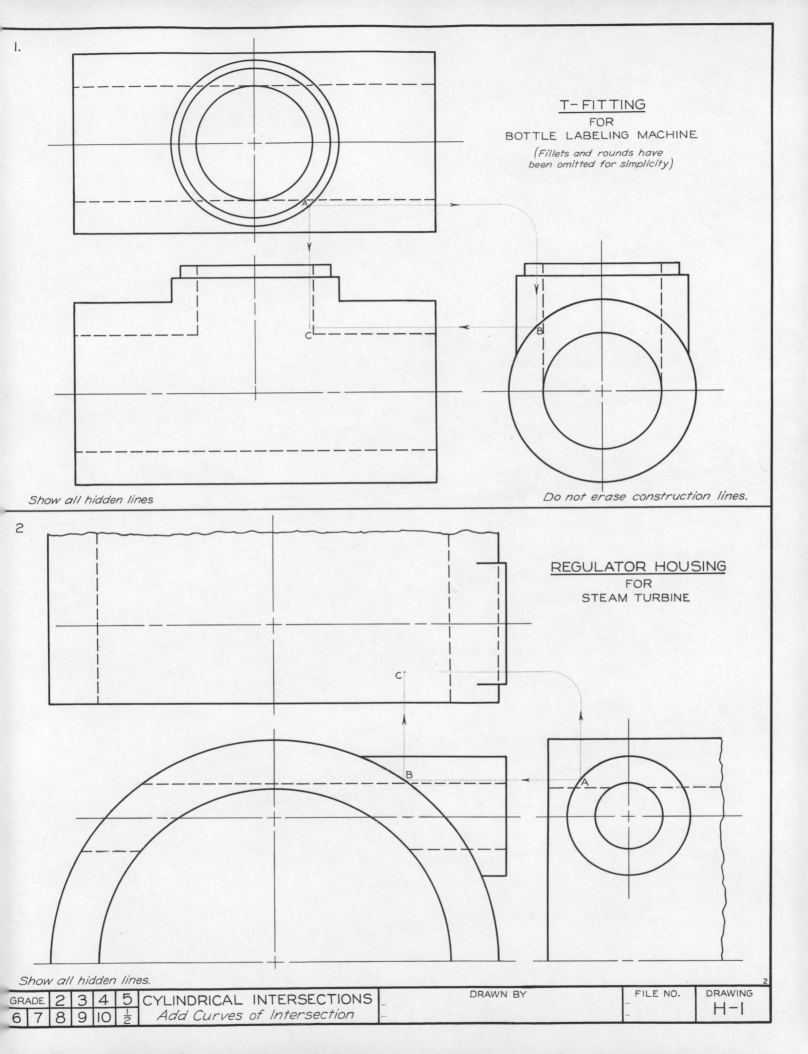

1.

T-FITTING
FOR
BOTTLE LABELING MACHINE

*(Fillets and rounds have
been omitted for simplicity)*

A

C B

Show all hidden lines

Do not erase construction lines.

2

REGULATOR HOUSING
FOR
STEAM TURBINE

C

B A

Show all hidden lines.

GRADE	2	3	4	5	CYLINDRICAL INTERSECTIONS	DRAWN BY	FILE NO.	DRAWING
6	7	8	9	10	½	*Add Curves of Intersection*			H-1

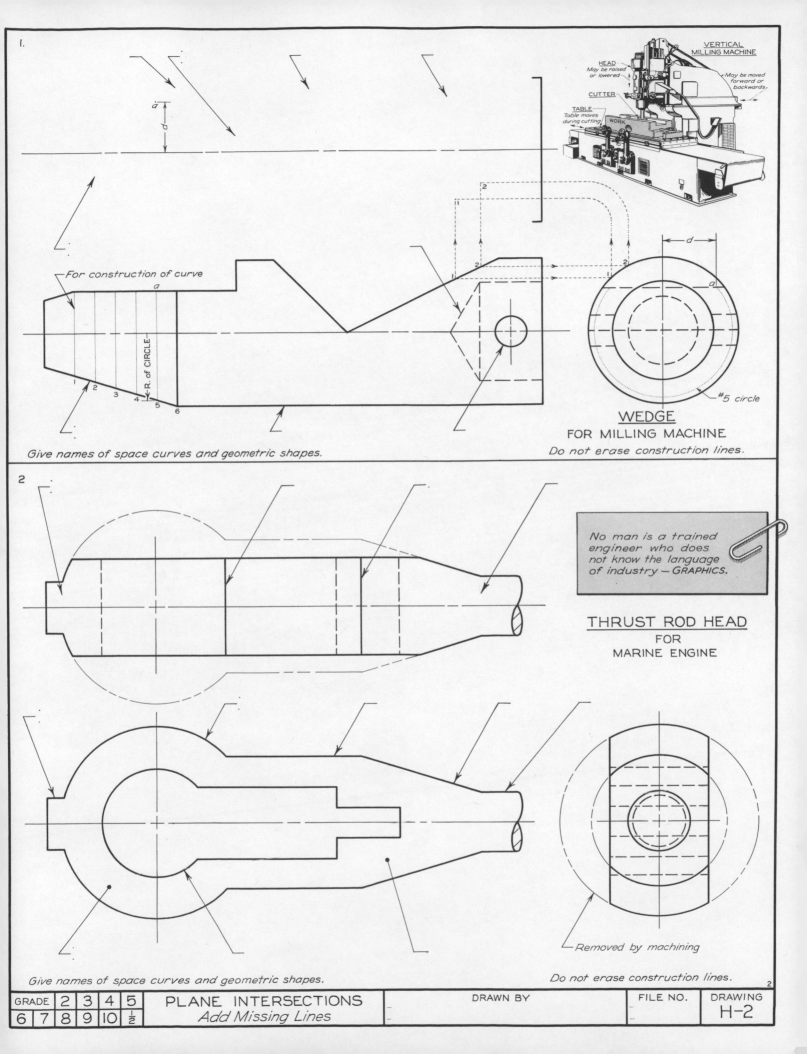

I.

VERTICAL
MILLING MACHINE

HEAD
May be raised
or lowered

May be moved
forward or
backwards

CUTTER

TABLE
Table moves
during cutting

WORK

For construction of curve

a

R. of CIRCLE

1 2 3 4 5 6

d

a

#5 circle

WEDGE
FOR MILLING MACHINE

Give names of space curves and geometric shapes.

Do not erase construction lines.

2.

No man is a trained
engineer who does
not know the language
of industry — GRAPHICS.

THRUST ROD HEAD
FOR
MARINE ENGINE

Removed by machining

Give names of space curves and geometric shapes.

Do not erase construction lines.

2

GRADE	2	3	4	5	PLANE INTERSECTIONS	DRAWN BY	FILE NO.	DRAWING
	6	7	8	9	10	½	*Add Missing Lines*			H-2

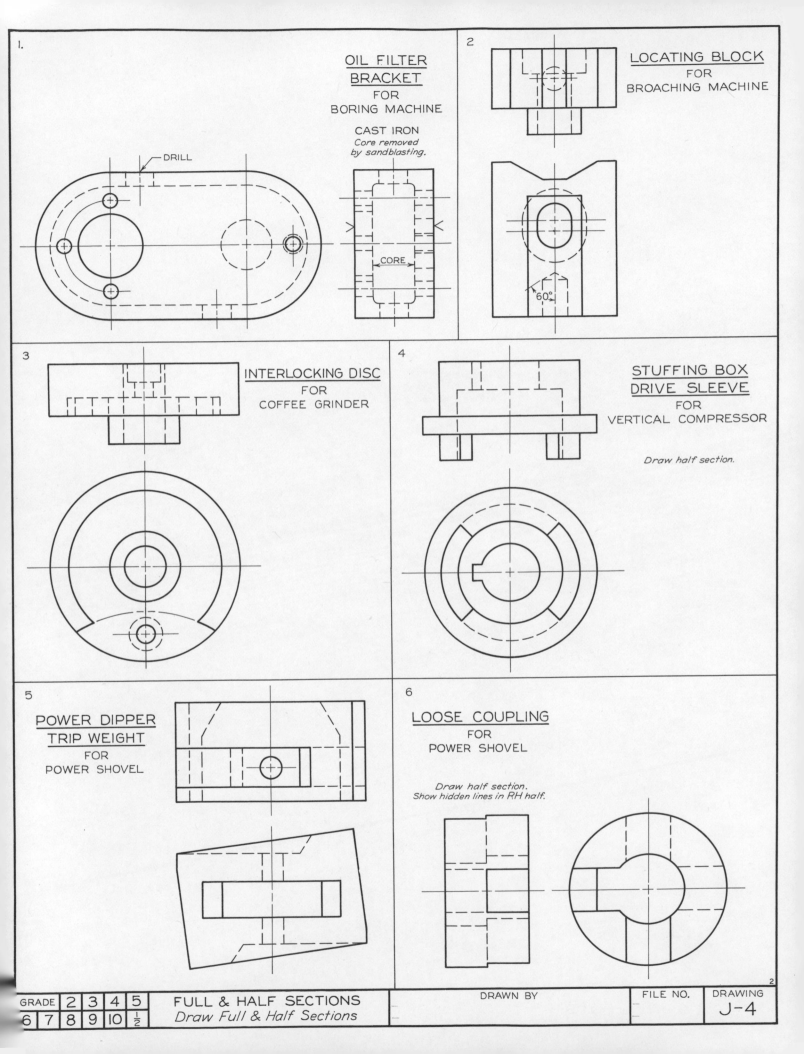

1. **OIL FILTER BRACKET** FOR BORING MACHINE

CAST IRON
Core removed by sandblasting.

DRILL

CORE

2. **LOCATING BLOCK** FOR BROACHING MACHINE

60°

3. **INTERLOCKING DISC** FOR COFFEE GRINDER

4. **STUFFING BOX DRIVE SLEEVE** FOR VERTICAL COMPRESSOR

Draw half section.

5. **POWER DIPPER TRIP WEIGHT** FOR POWER SHOVEL

6. **LOOSE COUPLING** FOR POWER SHOVEL

Draw half section.
Show hidden lines in RH half.

GRADE	2	3	4	5
6	7	8	9	10	½

FULL & HALF SECTIONS
Draw Full & Half Sections

DRAWN BY

FILE NO.

DRAWING
J-4

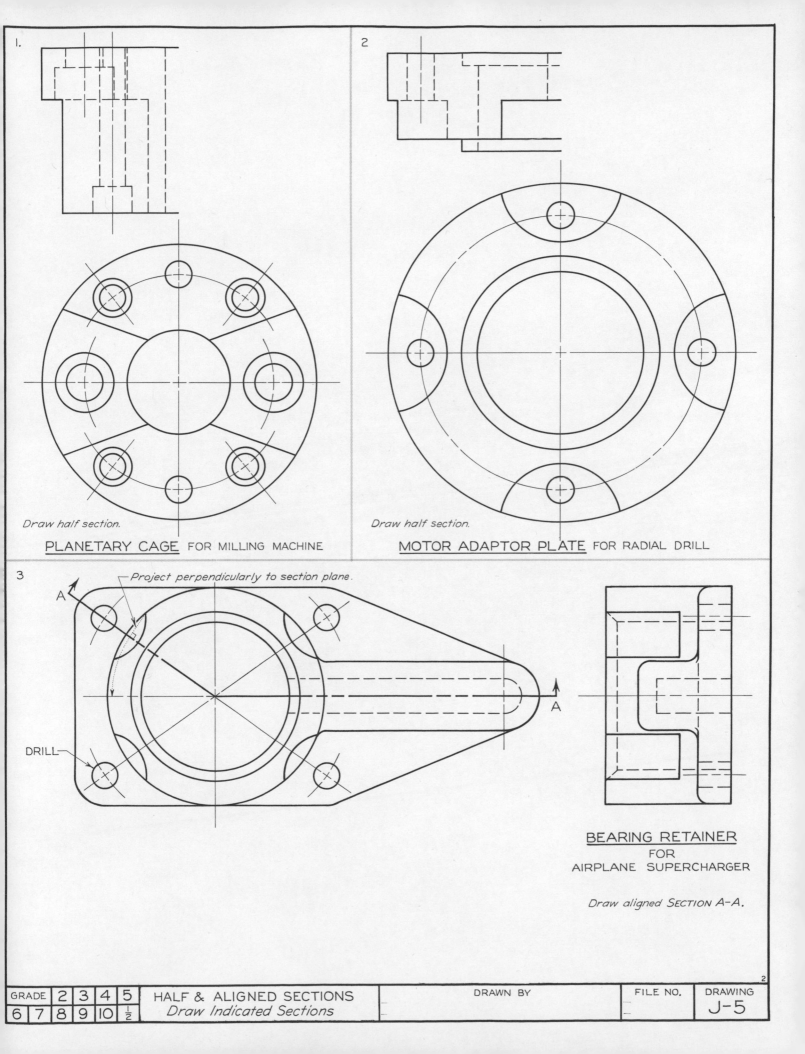

1.

Draw half section.

PLANETARY CAGE FOR MILLING MACHINE

2

Draw half section.

MOTOR ADAPTOR PLATE FOR RADIAL DRILL

3

Project perpendicularly to section plane.

A

A

DRILL

BEARING RETAINER
FOR
AIRPLANE SUPERCHARGER

Draw aligned SECTION A-A.

GRADE	2	3	4	5	HALF & ALIGNED SECTIONS	DRAWN BY	FILE NO.	DRAWING
6	7	8	9	10	½	*Draw Indicated Sections*			J-5

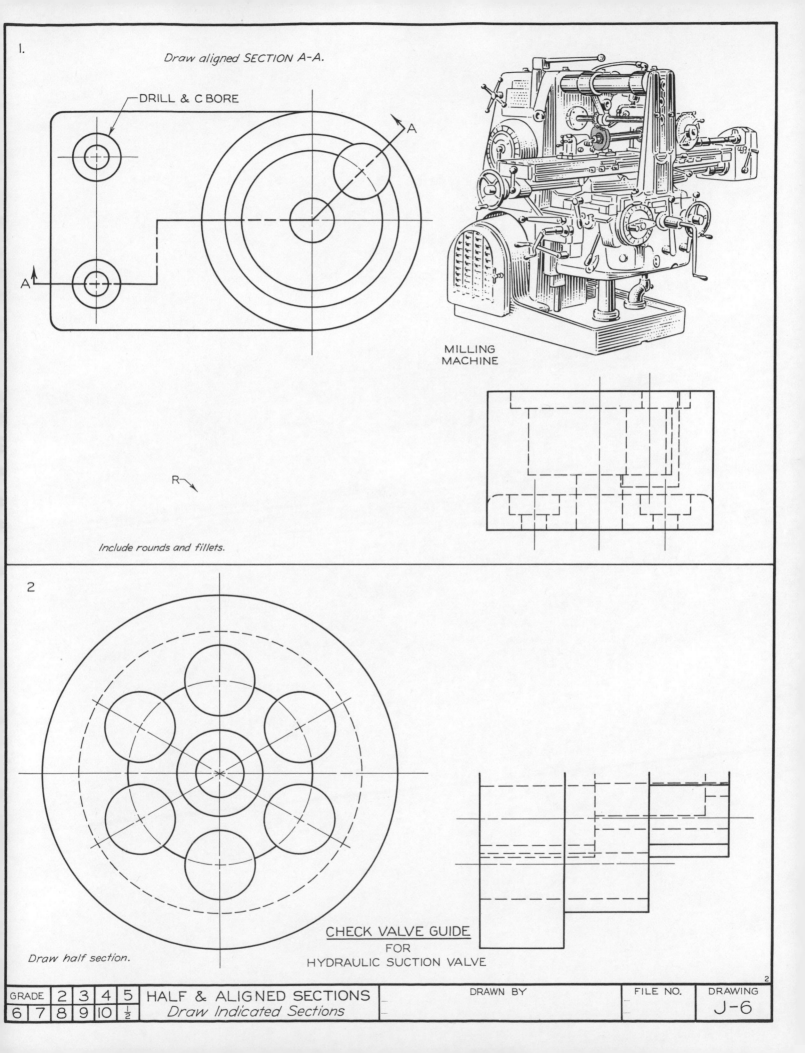

1.

Draw aligned SECTION A-A.

DRILL & C BORE

A

A

R

Include rounds and fillets.

MILLING
MACHINE

2

Draw half section.

CHECK VALVE GUIDE
FOR
HYDRAULIC SUCTION VALVE

GRADE	2	3	4	5	HALF & ALIGNED SECTIONS	DRAWN BY	FILE NO.	DRAWING
	6	7	8	9	10	½	*Draw Indicated Sections*			J-6

2

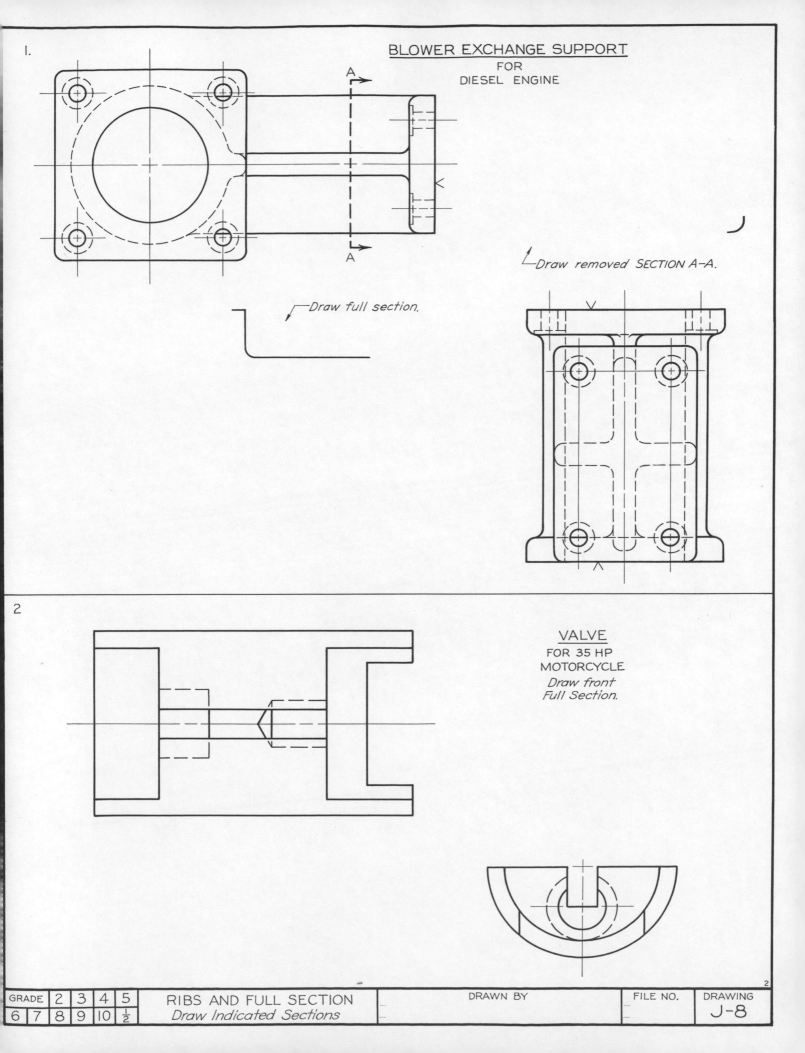

I.

BLOWER EXCHANGE SUPPORT
FOR
DIESEL ENGINE

A

A

Draw full section.

Draw removed SECTION A-A.

2

VALVE
FOR 35 HP
MOTORCYCLE
Draw front
Full Section.

GRADE	2	3	4	5
	6	7	8	9	10	½

RIBS AND FULL SECTION
Draw Indicated Sections

DRAWN BY

FILE NO.

DRAWING
J-8

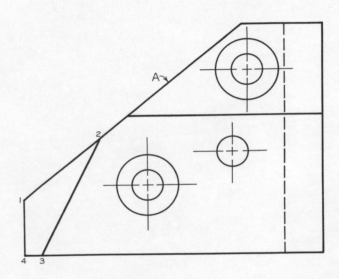

L
VIEW

To draw the side view of surface 1-2-3-4 it is necessary to have the surface as a line in an auxiliary view drawn first with the slope of the line according to its predetermined angle. In industrial practice it is frequently necessary to draw auxiliary views before regular views.

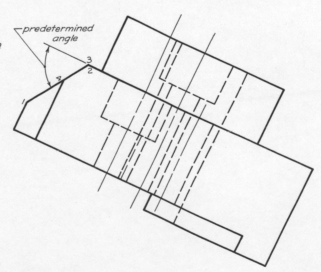

predetermined angle

INSTRUCTIONS

1- Draw left side view.

2- Draw complete auxiliary view showing true size and shape of surface A.

GRADE	2	3	4	5	AUXILIARY VIEWS	DRAWN BY	FILE NO.	DRAWING
6	7	8	9	10	½	*Draw Required Views*			K-4

I.

ADJUSTABLE CLAMP
FOR MILLING FIXTURE

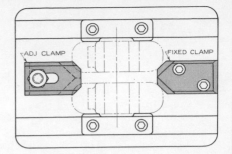

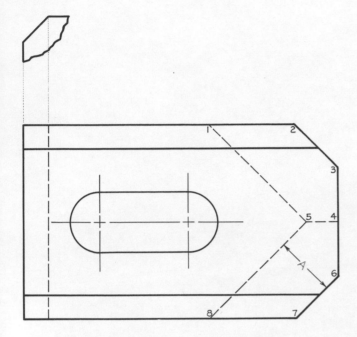

Draw complete auxiliary view showing surface A as a line and it's angle with the rear face. Letter the 8 numbers in auxiliary view.

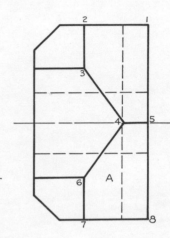

2

Draw complete auxiliary view showing surfaces A and B as lines and the true angle between them.

FIXED CLAMP
FOR MILLING FIXTURE

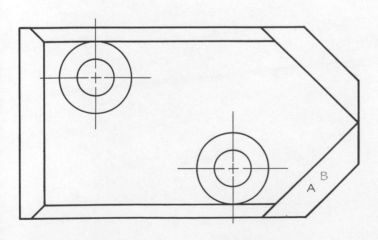

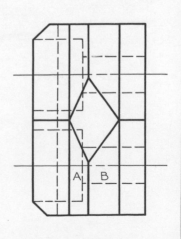

GRADE	2	3	4	5	AUXILIARY VIEWS	DRAWN BY	FILE NO.	DRAWING
6	7	8	9	10	½	Draw Required Views			K-5

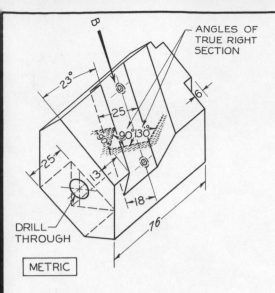

ANGLES OF
TRUE RIGHT
SECTION

23°

25

6

90° 130°

25

13

18

DRILL
THROUGH

76

METRIC

45° REAR TOOL BLADE HOLDER.

Draw:
(a) Partial depth auxiliary view in
direction indicated by arrow A.
(b) Depth auxiliary view followed
by secondary auxiliary view of
entire object, showing true right
section of groove as seen in
direction of arrow B, above.

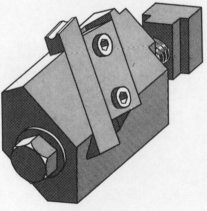

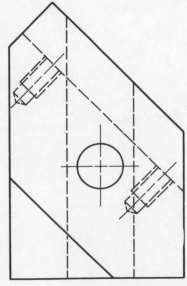

A

SCALE – FULL SIZE

GRADE	2	3	4	5	SECONDARY AUXILIARY VIEW	DRAWN BY		FILE NO.	DRAWING
6	7	8	9	10	½	*Draw Assigned Views.*			K-7

I.

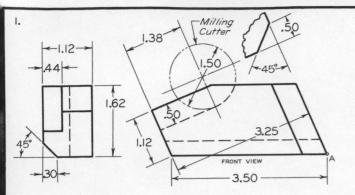

1.38

Milling Cutter

1.50

.50

45°

1.12

.44

1.62

45°

.30

.50

3.25

1.12

FRONT VIEW

A

3.50

Draw front view revolved clock-
wise on A as a pivot until milled
slot is horizontal for cutting on
a milling machine. Then draw
top and left side views.
In industry a milling fixture
is designed to hold the part in
this rotated position.

CAM AND CRANK
BORING TOOL
FOR
MILLING MACHINE

°A

2.

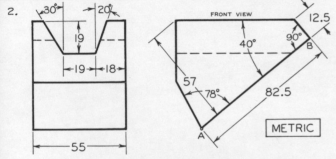

30° 20°

19

19 18

FRONT VIEW

12.5

40° 90° B

57

78°

82.5

METRIC

A

55

Draw front view revolved counterclockwise on A as a pivot
until AB is horizontal for plain milling of surface AB.
Then draw top and left side views.

WEDGE BLOCK
FOR
POWER SHOVEL

°A

GRADE	2	3	4	5	REVOLUTION	DRAWN BY	FILE NO.	DRAWING
	6	7	8	9	10	½	*Draw Revolved Views as Indicated*			L-1

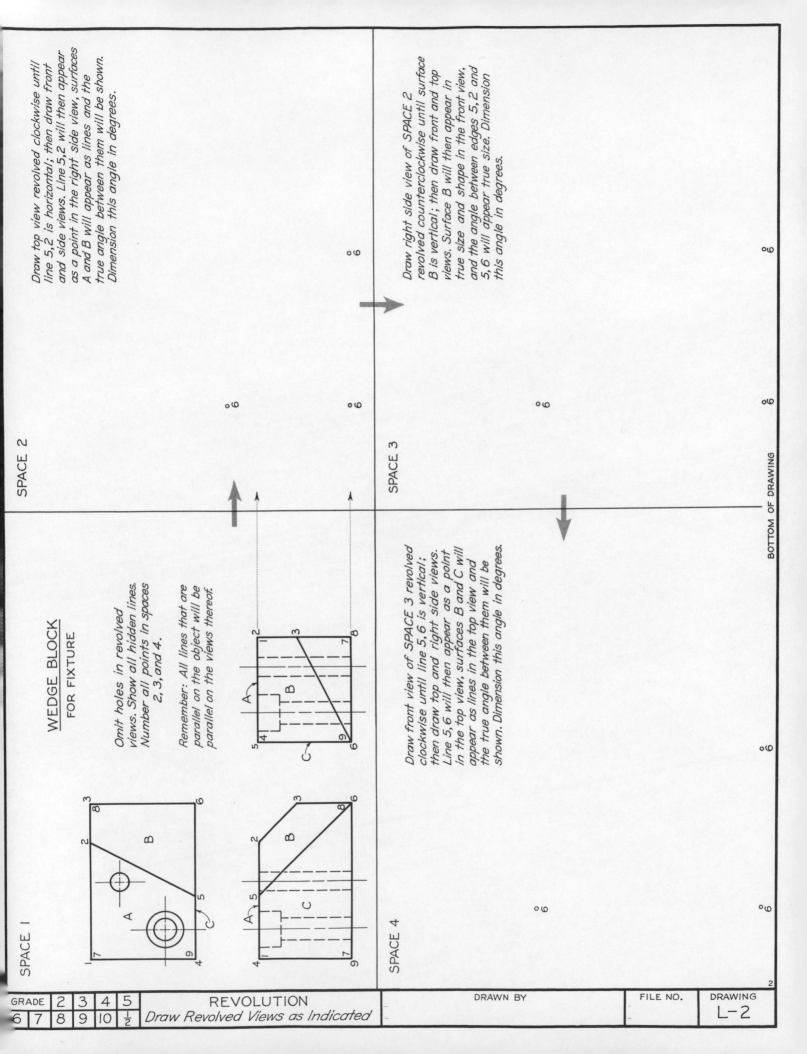

SPACE 2

Draw top view revolved clockwise until line 5,2 is horizontal; then draw front and side views. Line 5,2 will then appear as a point in the right side view, surfaces A and B will appear as lines and the true angle between them will be shown. Dimension this angle in degrees.

SPACE 3

Draw right side view of SPACE 2 revolved counterclockwise until surface B is vertical; then draw front and top views. Surface B will then appear in true size and shape in the front view, and the angle between edges 5,2 and 5,6 will appear true size. Dimension this angle in degrees.

SPACE 1

WEDGE BLOCK
FOR FIXTURE

Omit holes in revolved views. Show all hidden lines. Number all points in spaces 2, 3, and 4.

Remember: All lines that are parallel on the object will be parallel on the views thereof.

SPACE 4

Draw front view of SPACE 3 revolved clockwise until line 5,6 is vertical; then draw top and right side views. Line 5,6 will then appear as a point in the top view, surfaces B and C will appear as lines in the top view and the true angle between them will be shown. Dimension this angle in degrees.

BOTTOM OF DRAWING

GRADE	2	3	4	5
6	7	8	9	10	½

REVOLUTION
Draw Revolved Views as Indicated

DRAWN BY

FILE NO.

DRAWING
L-2

NORMAL SURFACES (All)

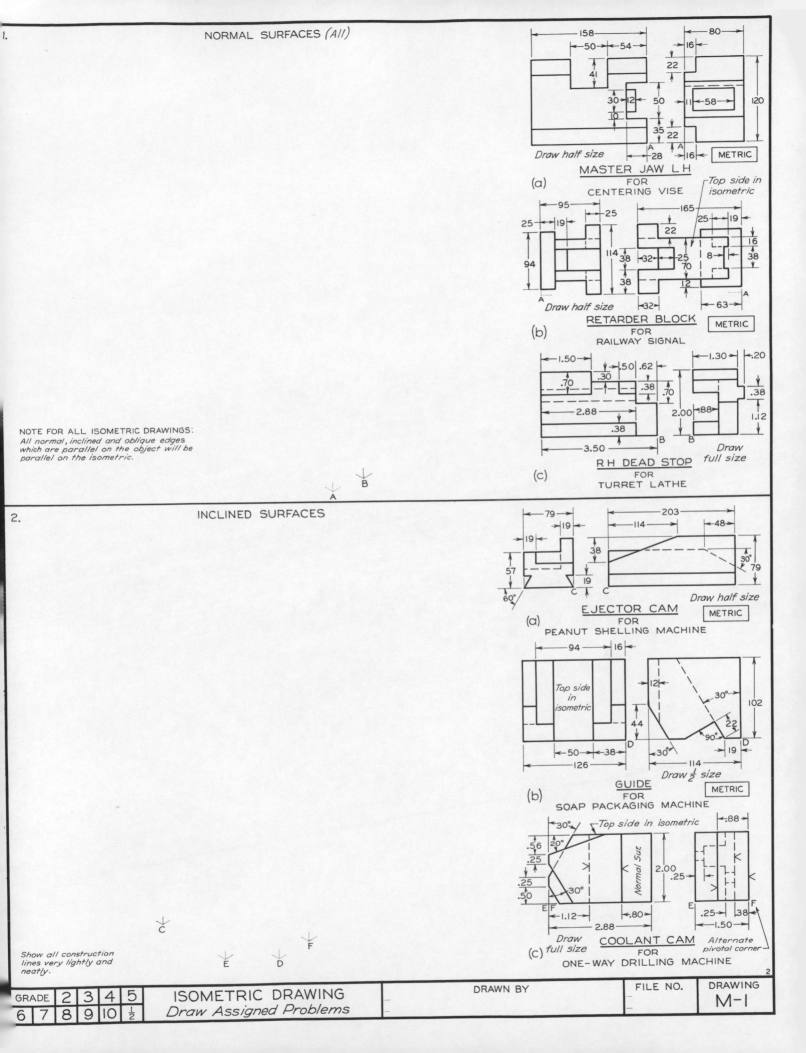

MASTER JAW LH
FOR
CENTERING VISE

(a)

158 · 50 · 54 · 80 · 16 · 41 · 22 · 30 · 12 · 50 · 11 · 58 · 120 · 10 · 35 · 22

Draw half size

28 · 16

METRIC

Top side in isometric

RETARDER BLOCK
FOR
RAILWAY SIGNAL

(b)

95 · 19 · 25 · 165 · 25 · 19 · 25 · 22 · 16 · 94 · 114 · 38 · 32 · 25 · 8 · 38 · 70 · 38 · 12 · 32 · 63

A   Draw half size   A

METRIC

NOTE FOR ALL ISOMETRIC DRAWINGS:
All normal, inclined and oblique edges
which are parallel on the object will be
parallel on the isometric.

R H DEAD STOP
FOR
TURRET LATHE

(c)

1.50 · .50 · .62 · .30 · .38 · .70 · .70 · 2.88 · 2.00 · .88 · .38 · .38 · 1.12 · 3.50   B   B · 1.30 · .20

Draw full size

B
A

INCLINED SURFACES

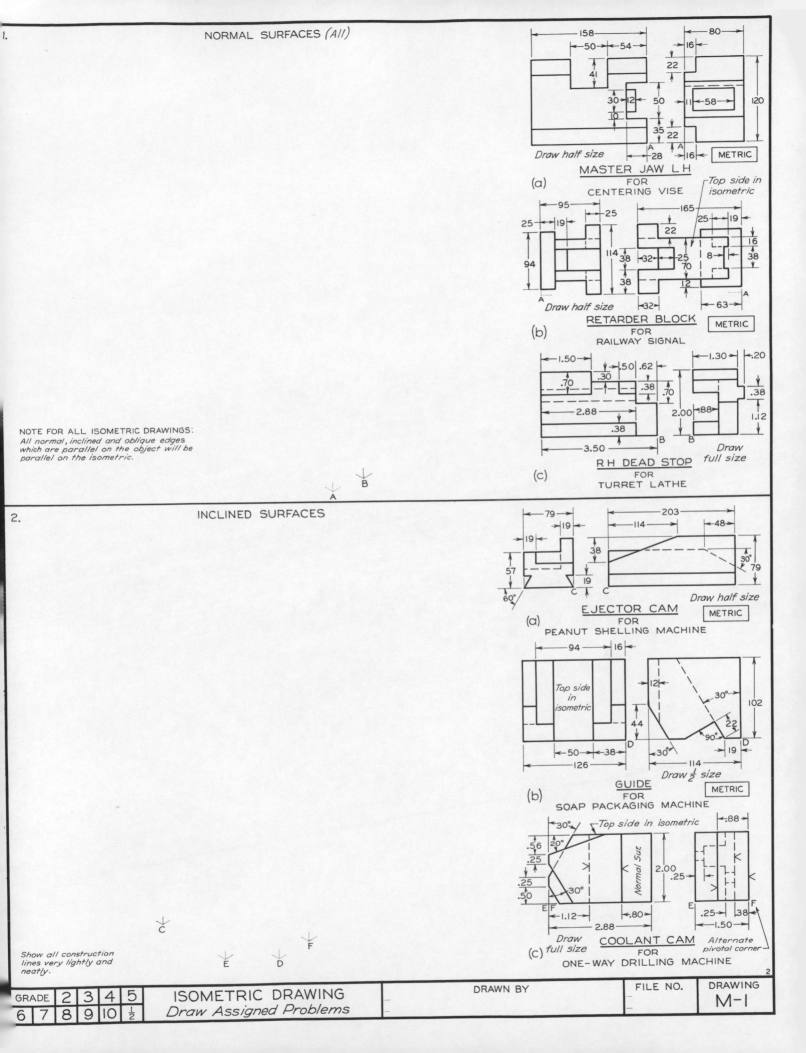

EJECTOR CAM
FOR
PEANUT SHELLING MACHINE

(a)

79 · 19 · 19 · 203 · 114 · 48 · 57 · 38 · 30° · 79 · 19 · 60°   C

Draw half size

METRIC

GUIDE
FOR
SOAP PACKAGING MACHINE

(b)

94 · 16 · Top side in isometric · 12 · 30° · 102 · 44 · 22 · 90° · 50 · 38 · D · 30° · 114 · 19 · D · 126

Draw ½ size

METRIC

COOLANT CAM
FOR
ONE-WAY DRILLING MACHINE

(c)

30° · Top side in isometric · .88 · .56 · 20° · .25 · Normal Sur. · 2.00 · .25 · .25 · .50 · 30° · E F · 1.12 · .80 · E · .25 · .38 · F · 2.88 · 1.50

Draw full size

Alternate pivotal corner

Show all construction
lines very lightly and
neatly.

C   F   E   D

2

GRADE	2	3	4	5
	6	7	8	9	10	½

ISOMETRIC DRAWING
Draw Assigned Problems

DRAWN BY

FILE NO.

DRAWING
M-1

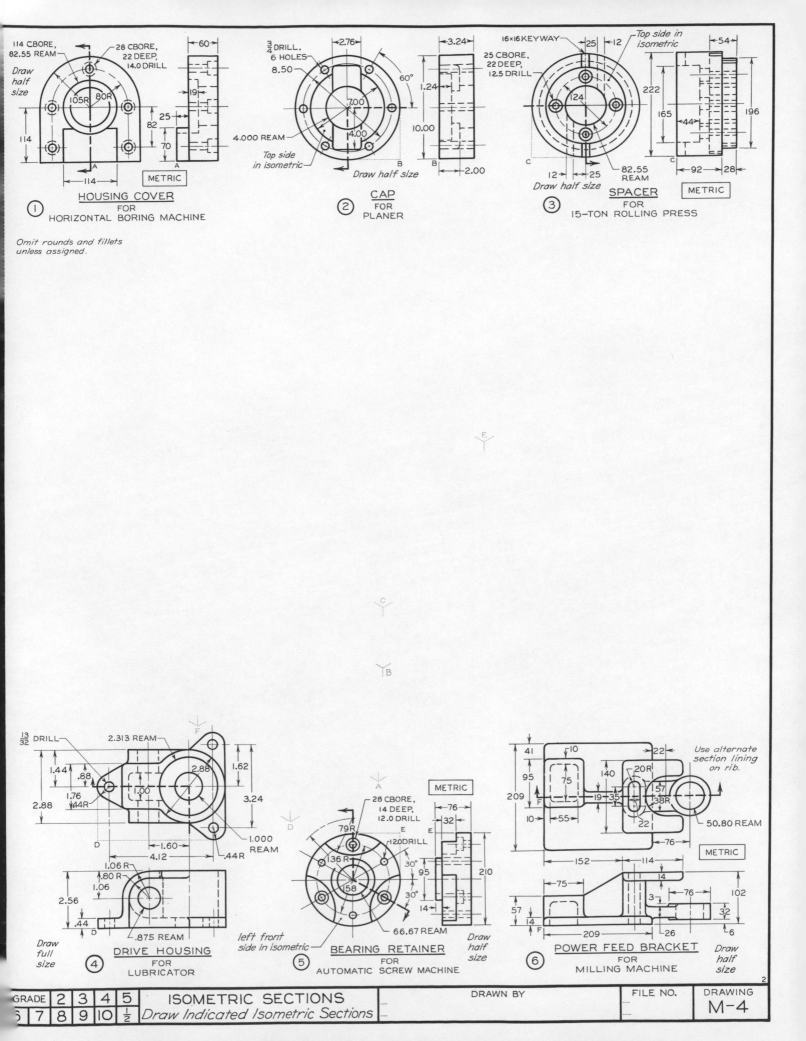

① HOUSING COVER
FOR
HORIZONTAL BORING MACHINE

114 CBORE,
82.55 REAM
28 CBORE,
22 DEEP, 14.0 DRILL
Draw half size
60
19
105R 80R
25
82
70
114
114
A
A
METRIC

Omit rounds and fillets
unless assigned.

② CAP
FOR
PLANER

¾ DRILL, 6 HOLES
8.50
2.76
60°
7.00
4.00
3.24
1.24
10.00
4.000 REAM
Top side in isometric
Draw half size
B
B
2.00

③ SPACER
FOR
15-TON ROLLING PRESS

16×16 KEYWAY
Top side in isometric
25 12
25 CBORE, 22 DEEP, 12.5 DRILL
124
222
165
196
44
54
12 25
82.55 REAM
Draw half size
c
c
92 28
METRIC

④ DRIVE HOUSING
FOR
LUBRICATOR

13/32 DRILL
2.313 REAM
F
1.44
.88
1.76
44R
1.00
2.88
2.88
1.62
3.24
2.88
1.000 REAM
.44R
D
1.60
4.12
D
1.06 R
.80 R
1.06
2.56
.44
D
.875 REAM
Draw full size

⑤ BEARING RETAINER
FOR
AUTOMATIC SCREW MACHINE

A
28 CBORE, 14 DEEP, 12.0 DRILL
METRIC
79R
136 R
120 DRILL
158
30°
30°
66.67 REAM
left front side in isometric
E
E
76
32
95
210
14
Draw half size

⑥ POWER FEED BRACKET
FOR
MILLING MACHINE

Use alternate section lining on rib.
41
10
22
95
75
140
20R
209
19 35
57
38R
10
55
22
50.80 REAM
152
114
METRIC
75
14
3
76
57
102
14
32
209
26
6
F
F
Draw half size

| GRADE | 2 | 3 | 4 | 5 | ISOMETRIC SECTIONS | DRAWN BY | FILE NO. | DRAWING |
| 6 | 7 | 8 | 9 | 10 ½ | Draw Indicated Isometric Sections | | | M-4 |

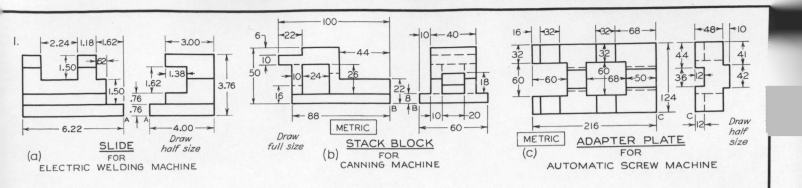

1.

SLIDE
FOR
ELECTRIC WELDING MACHINE

(a)

Draw half size

STACK BLOCK
FOR
CANNING MACHINE

(b)

METRIC

Draw full size

ADAPTER PLATE
FOR
AUTOMATIC SCREW MACHINE

(c)

METRIC

Draw half size

Draw all three problems in Cavalier projection. All surfaces are normal surfaces.

2

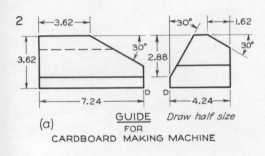

GUIDE
FOR
CARDBOARD MAKING MACHINE

(a)

Draw half size

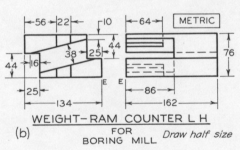

WEIGHT-RAM COUNTER L H
FOR
BORING MILL

(b)

METRIC

Draw half size

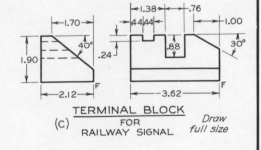

TERMINAL BLOCK
FOR
RAILWAY SIGNAL

(c)

Draw full size

Draw all three problems in Cabinet projection.

2

GRADE	2	3	4	5	OBLIQUE PROJECTION	DRAWN BY	FILE NO.	DRAWING
6	7	8	9	10	½	*Draw Assigned Problems*			N-1

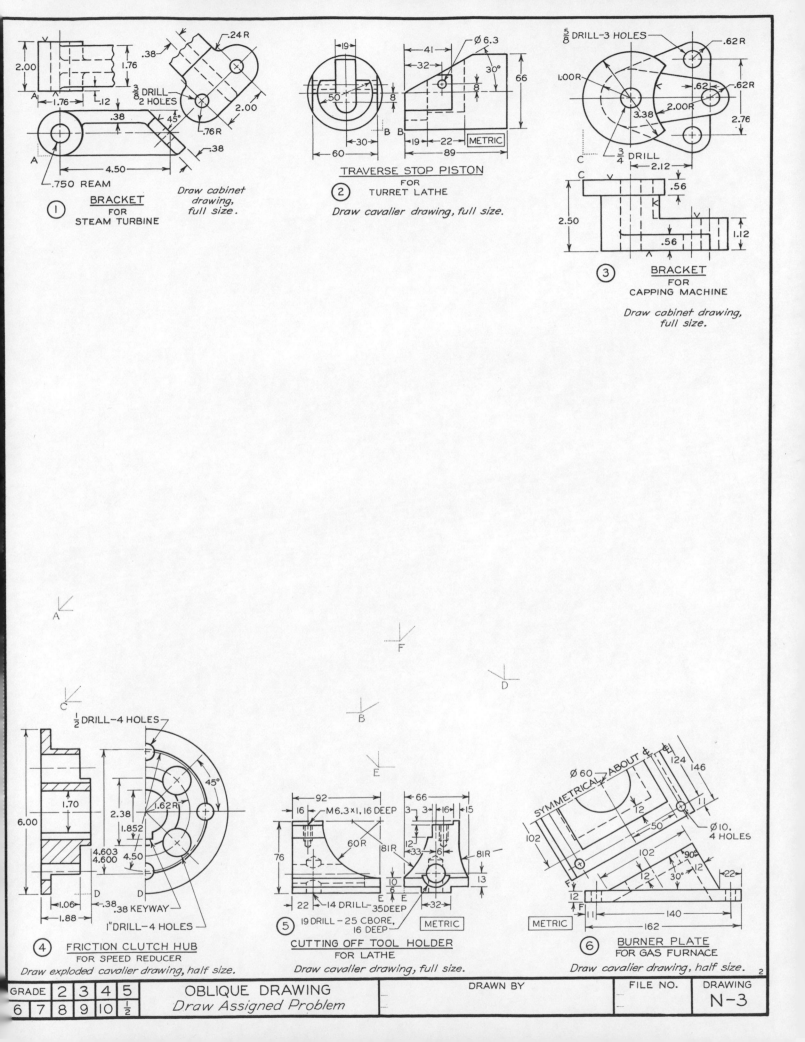

1 BRACKET
FOR
STEAM TURBINE

2.00 1.76 .38 .24 R

1.76 .12 3/8 DRILL 2 HOLES

.38 45° .76 R 2.00

4.50 .38

.750 REAM

Draw cabinet drawing, full size.

TRAVERSE STOP PISTON
2 FOR
TURRET LATHE

19 41 Ø 6.3

32 30° 66

50 8

30 B B

60 19 22 METRIC 8

89

Draw cavalier drawing, full size.

3 BRACKET
FOR
CAPPING MACHINE

5/8 DRILL-3 HOLES .62 R

1.00R .62 .62R

3.38 2.00R 2.76

C 3/4 DRILL 2.12

C .56

2.50 K

.56 1.12

Draw cabinet drawing, full size.

A

F

D

B

E

C

4 FRICTION CLUTCH HUB
FOR SPEED REDUCER

1/2 DRILL-4 HOLES

F 45°

1.70 2.38 1.62R

1.852

6.00 4.603 4.50
4.600

1.06 D D

1.88 .38 .38 KEYWAY

1"DRILL-4 HOLES

Draw exploded cavalier drawing, half size.

5 CUTTING OFF TOOL HOLDER
FOR LATHE

92 66

16 M6.3×1,16 DEEP 3 3 16 15

60R 12

76 81R 33 6 81R

10 13
6

22 14 DRILL E E 32
35DEEP

19 DRILL - 25 CBORE, METRIC
16 DEEP

Draw cavalier drawing, full size.

6 BURNER PLATE
FOR GAS FURNACE

Ø 60 SYMMETRICAL ABOUT ℄ 124 146

12 11

102 50 Ø 10,
4 HOLES

102 90°

12 30° 12

F 12 22

METRIC 12 11 140

F 162

Draw cavalier drawing, half size.

2

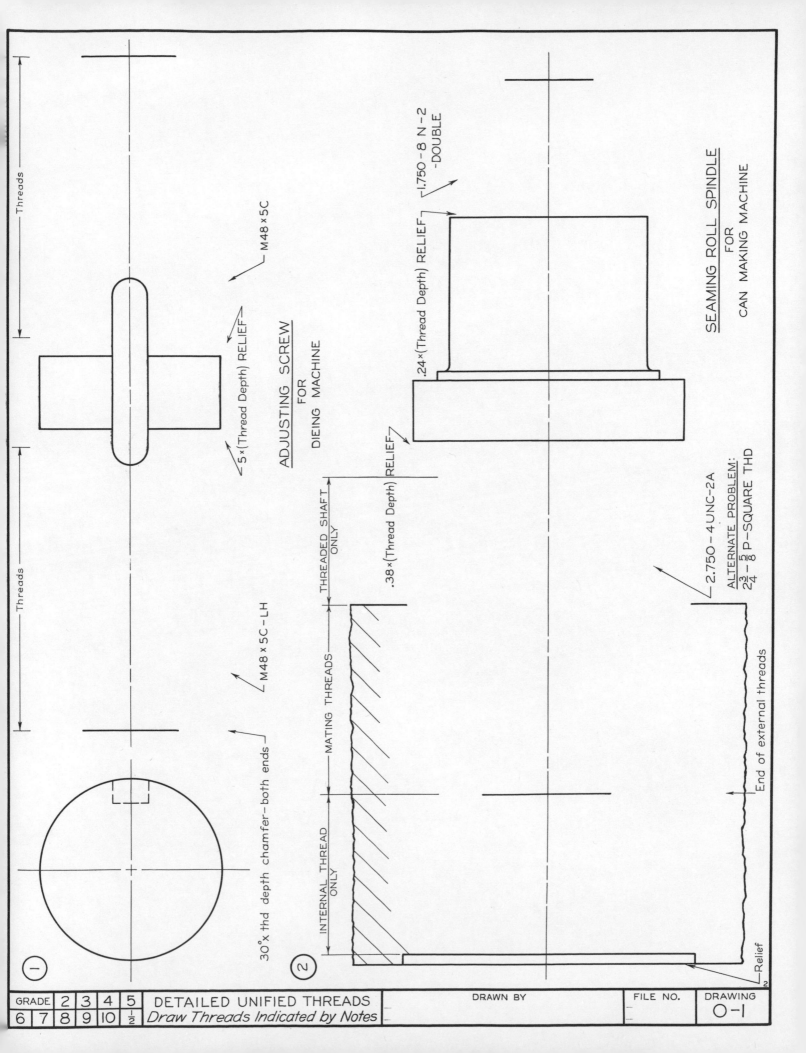

Threads

Threads

M48 × 5C

↙ 5 × (Thread Depth) RELIEF →

ADJUSTING SCREW
FOR
DIEING MACHINE

↙ M48 × 5C − LH

30° × thd depth chamfer–both ends

① −

1.750 − 8 N − 2
−DOUBLE

.24 × (Thread Depth) RELIEF

SEAMING ROLL SPINDLE
FOR
CAN MAKING MACHINE

THREADED SHAFT
ONLY

.38 × (Thread Depth) RELIEF

MATING THREADS

INTERNAL THREAD
ONLY

②

2.750 − 4 UNC − 2A

ALTERNATE PROBLEM:
$2\frac{3}{4} - \frac{5}{8}$ P−SQUARE THD

End of external threads

Relief

2

GRADE	2	3	4	5	DETAILED UNIFIED THREADS	DRAWN BY	FILE NO.	DRAWING
	6	7	8	9	10	$\frac{1}{2}$	*Draw Threads Indicated by Notes*			O−1

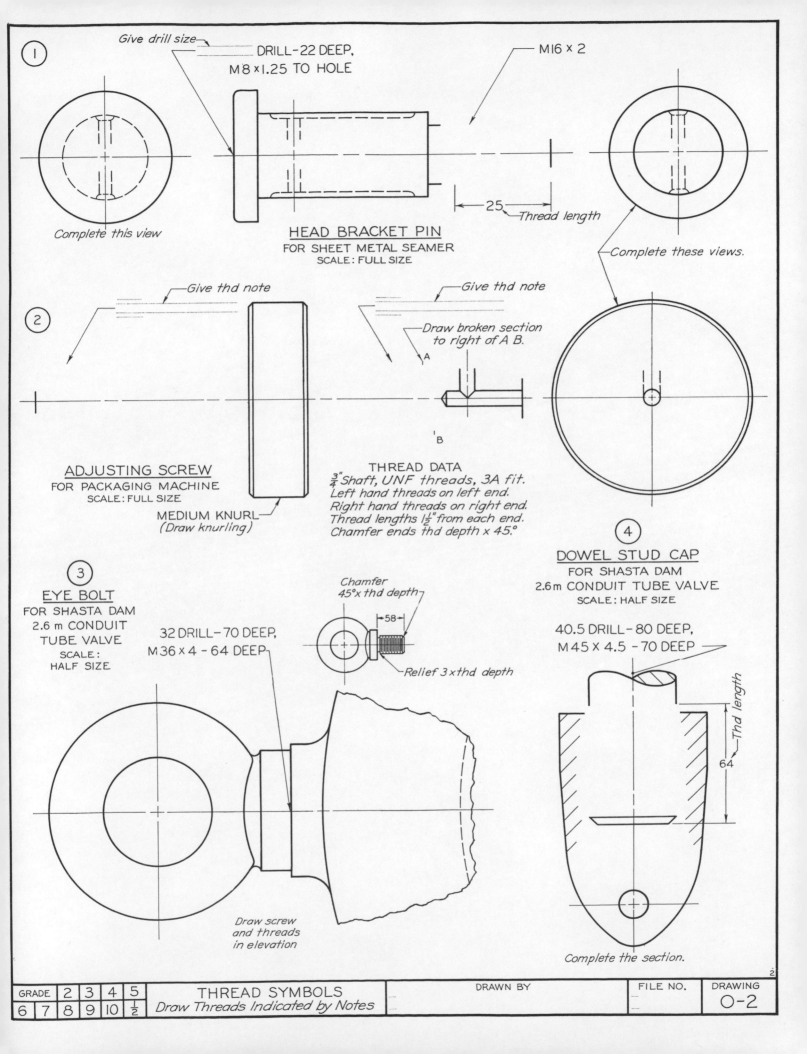

① Give drill size — DRILL-22 DEEP, M8×1.25 TO HOLE

M16 × 2

Complete this view

HEAD BRACKET PIN
FOR SHEET METAL SEAMER
SCALE: FULL SIZE

25 — Thread length

Complete these views.

② Give thd note

Give thd note

Draw broken section to right of A B.

A

B

ADJUSTING SCREW
FOR PACKAGING MACHINE
SCALE: FULL SIZE

MEDIUM KNURL
(Draw knurling)

THREAD DATA
$\frac{3}{4}$" Shaft, UNF threads, 3A fit.
Left hand threads on left end.
Right hand threads on right end.
Thread lengths $1\frac{1}{2}$" from each end.
Chamfer ends thd depth × 45.°

④
DOWEL STUD CAP
FOR SHASTA DAM
2.6 m CONDUIT TUBE VALVE
SCALE: HALF SIZE

③
EYE BOLT
FOR SHASTA DAM
2.6 m CONDUIT
TUBE VALVE
SCALE:
HALF SIZE

32 DRILL-70 DEEP,
M36 × 4 - 64 DEEP

Chamfer
45°× thd depth

58

Relief 3 × thd depth

Draw screw
and threads
in elevation

40.5 DRILL - 80 DEEP,
M45 × 4.5 - 70 DEEP

Thd length

64

Complete the section.

GRADE	2	3	4	5	THREAD SYMBOLS	DRAWN BY	FILE NO.	DRAWING
6	7	8	9	10	$\frac{1}{2}$	*Draw Threads Indicated by Notes*			O-2

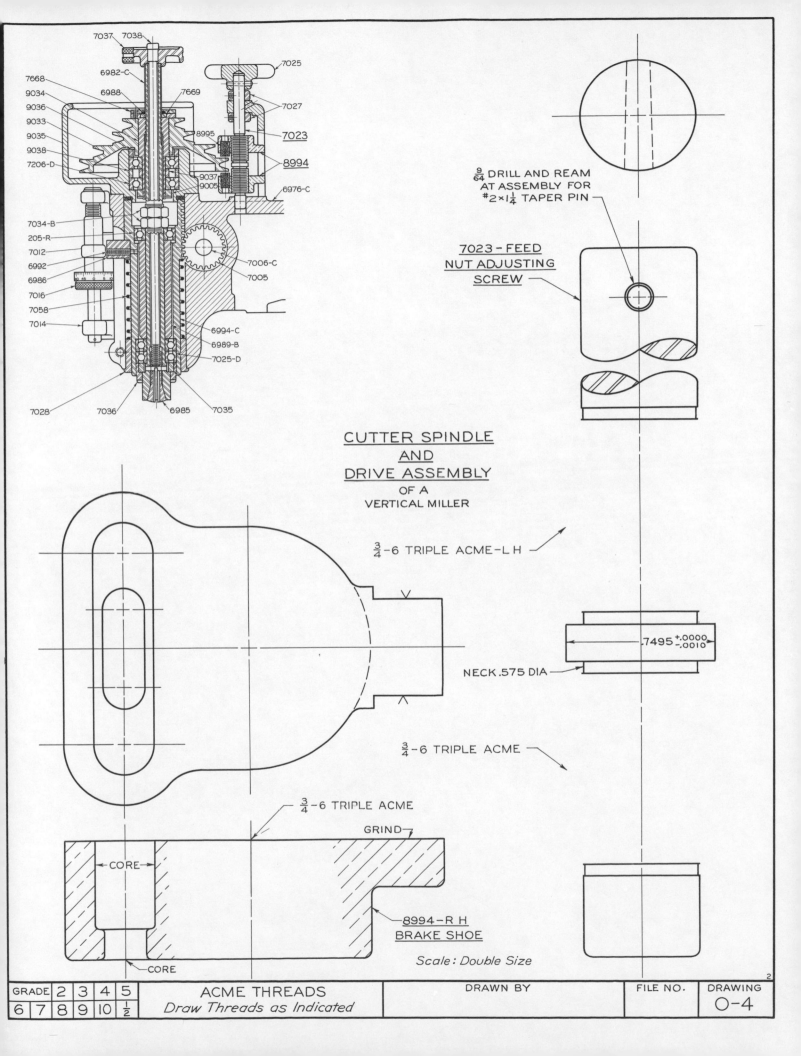

7037 7038

6982-C

7668
9034
9036
9033
9035
9038
7206-D

6988

7669

7025

7027

7023

8994

8995

9037
9005

6976-C

7034-B
205-R
7012
6992
6986
7016
7058
7014

7006-C
7005

6994-C
6989-B
7025-D

7028   7036   6985   7035

$\frac{9}{64}$ DRILL AND REAM
AT ASSEMBLY FOR
#2×1$\frac{1}{4}$ TAPER PIN

7023 – FEED
NUT ADJUSTING
SCREW

CUTTER SPINDLE
AND
DRIVE ASSEMBLY
OF A
VERTICAL MILLER

$\frac{3}{4}$ – 6 TRIPLE ACME – L H

.7495 $^{+.0000}_{-.0010}$

NECK .575 DIA

$\frac{3}{4}$ – 6 TRIPLE ACME

$\frac{3}{4}$ – 6 TRIPLE ACME

GRIND

CORE

8994 – R H
BRAKE SHOE

CORE

Scale : Double Size

GRADE	2	3	4	5	ACME THREADS	DRAWN BY	FILE NO.	DRAWING
6	7	8	9	10	$\frac{1}{2}$	*Draw Threads as Indicated*			O-4

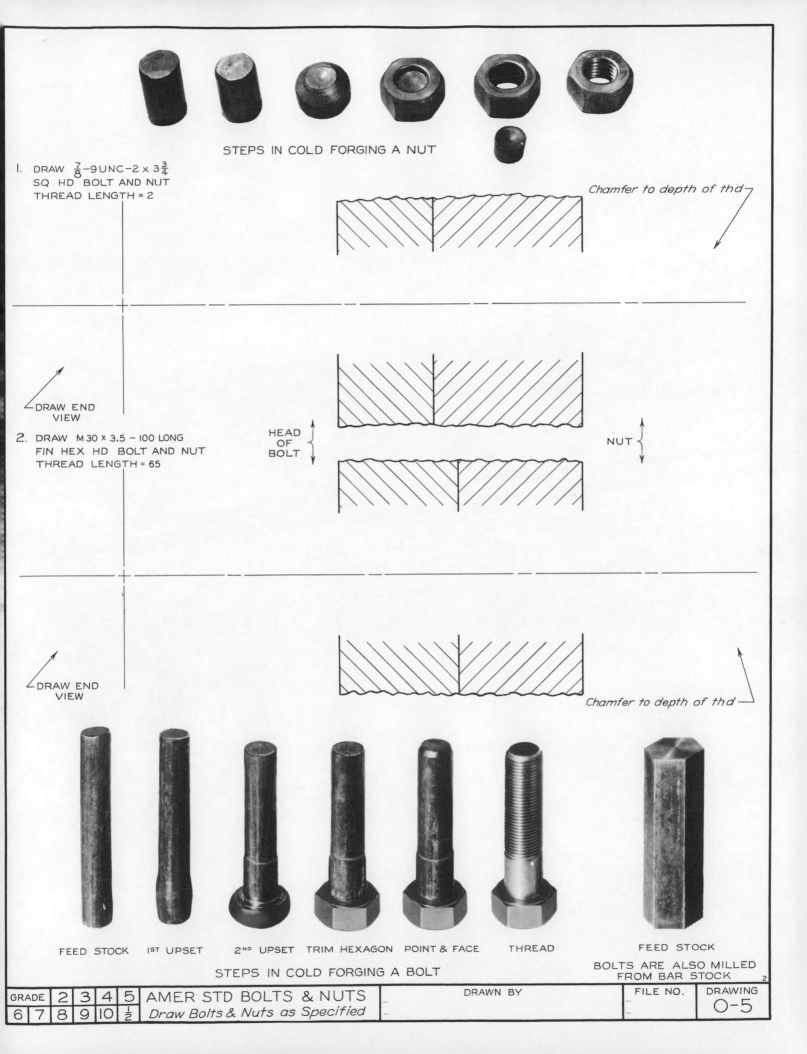

STEPS IN COLD FORGING A NUT

1. DRAW $\frac{7}{8}$–9UNC–2 × $3\frac{3}{4}$
   SQ HD BOLT AND NUT
   THREAD LENGTH = 2

Chamfer to depth of thd

DRAW END VIEW

2. DRAW M30 × 3.5 – 100 LONG
   FIN HEX HD BOLT AND NUT
   THREAD LENGTH = 65

HEAD OF BOLT

NUT

DRAW END VIEW

Chamfer to depth of thd

FEED STOCK   1ST UPSET   2ND UPSET   TRIM HEXAGON   POINT & FACE   THREAD   FEED STOCK

STEPS IN COLD FORGING A BOLT

BOLTS ARE ALSO MILLED FROM BAR STOCK

GRADE	2	3	4	5	AMER STD BOLTS & NUTS	DRAWN BY	FILE NO.	DRAWING
6	7	8	9	10	$\frac{1}{2}$	Draw Bolts & Nuts as Specified			O-5

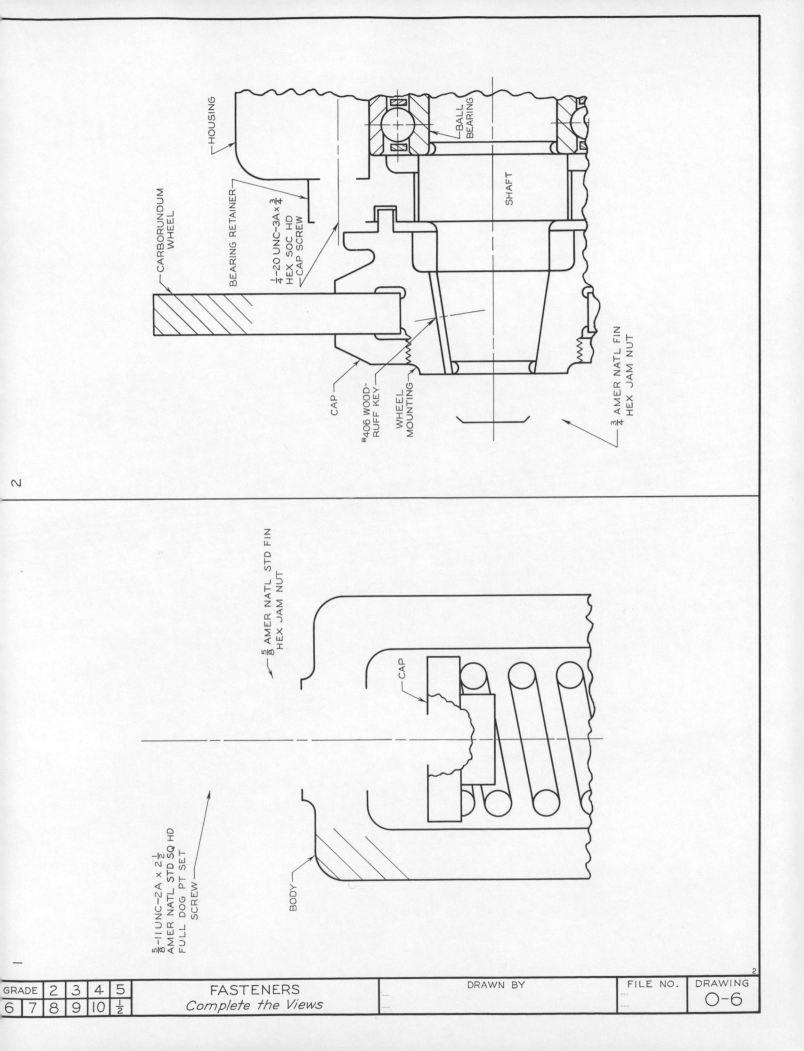

HOUSING

BALL
BEARING

CARBORUNDUM
WHEEL

BEARING RETAINER

SHAFT

$\frac{1}{4}$-20 UNC-3A x $\frac{3}{4}$
HEX SOC HD
CAP SCREW

CAP

#406 WOOD-
RUFF KEY

WHEEL
MOUNTING

$\frac{3}{4}$ AMER NATL FIN
HEX JAM NUT

$\frac{5}{8}$ AMER NATL STD FIN
HEX JAM NUT

CAP

$\frac{5}{8}$-11UNC-2A x $2\frac{1}{2}$
AMER NATL STD SQ HD
FULL DOG PT SET
SCREW

BODY

GRADE	2	3	4	5	FASTENERS	DRAWN BY	FILE NO.	DRAWING
6	7	8	9	10	$\frac{1}{2}$	*Complete the Views*			O-6

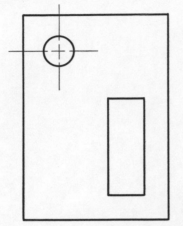

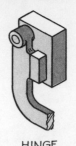

HINGE
BASE

SCALE: FULL SIZE

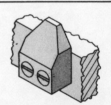

STOP
BLOCK

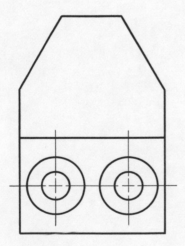

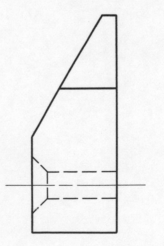

SCALE: HALF SIZE

GRADE	2	3	4	5	DIMENSIONING	DRAWN BY	FILE NO.	DRAWING
6	7	8	9	10	$\frac{1}{2}$	*Dimension the Views*			Q-1

1

Add two counterbored holes for cap screws in the views
of a key. Depth of counterbore is equal to height of cap
screw head. Draw diameters of drilled and counterbored
holes .04" larger than diameters of body and head of cap
screws. Dimension the views fully. Scale: Full size. F A O.

$\frac{5}{16}$ – 18 UNC–2A × 1$\frac{1}{2}$ HEX
SOC CAP SCREWS

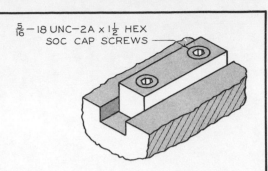

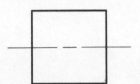

2

Add two drilled holes to the views of the packing gland,
allowing 1.0 for clearances. Allow 0.8 clearance on reamed
hole. Dimension the views fully, adding finish marks.
Scale: Half size.

METRIC

MIO × 1.5 × 70 HEX
HD CAP SCREWS

26 DIA

REAM

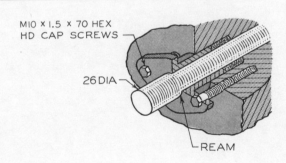

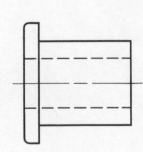

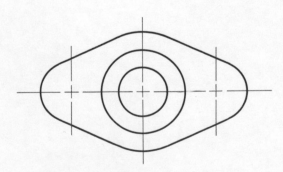

GRADE	2	3	4	5	DIMENSIONING	DRAWN BY	FILE NO.	DRAWING
	6	7	8	9	10	$\frac{1}{2}$	Dimension the Views			Q-2

1

Add 6 drilled holes 60° apart on given center lines, allowing 1.0 clearances for six M24 × 3 × 65 LG hex hd cap screws. Dimension the views of the Retaining Washer fully. The center hole is reamed to an RC 5 fit. Scale: One-fifth Size.

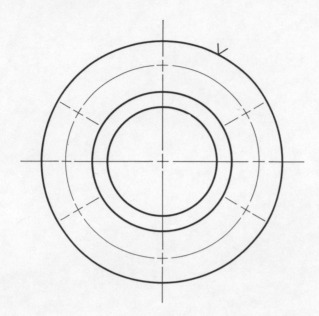 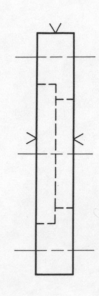

2

Dimension the views of the Eccentric fully. The hole is reamed to an RC 1 fit. Add finish marks. Scale: Half Size.

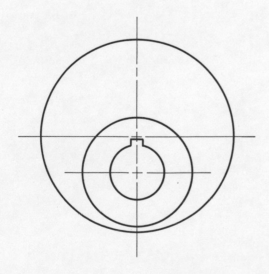 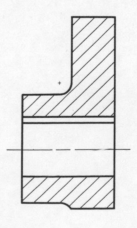

GRADE	2	3	4	5	DIMENSIONING	DRAWN BY	FILE NO.	DRAWING
6	7	8	9	10	½	Dimension the Views			Q-3

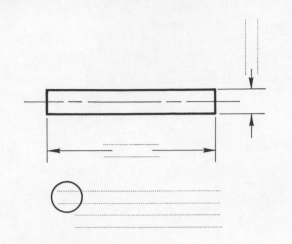

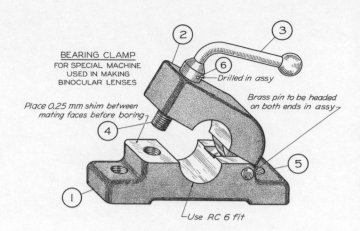

BEARING CLAMP
FOR SPECIAL MACHINE
USED IN MAKING
BINOCULAR LENSES

② ③ ⑥ *Drilled in assy*

Place 0.25 mm shim between mating faces before boring

④

Brass pin to be headed on both ends in assy

⑤

①

Use RC 6 fit

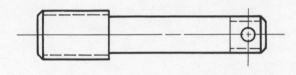

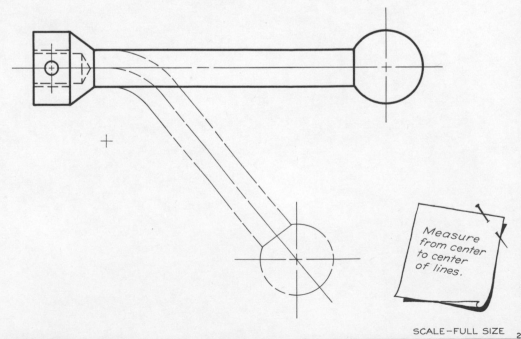

Measure from center to center of lines.

SCALE—FULL SIZE

GRADE	2	3	4	5	DIMENSIONING		DRAWN BY	FILE NO.	DRAWING
	6	7	8	9	10	½	*Add Complete Dimensions*		Q-5

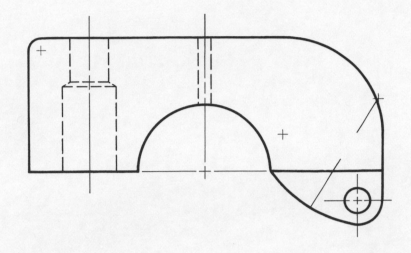

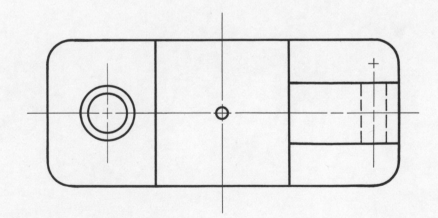

GRADE	2	3	4	5	DIMENSIONING	DRAWN BY	FILE NO.	DRAWING
	6	7	8	9	10	½	Add Complete Dimensions			Q-6

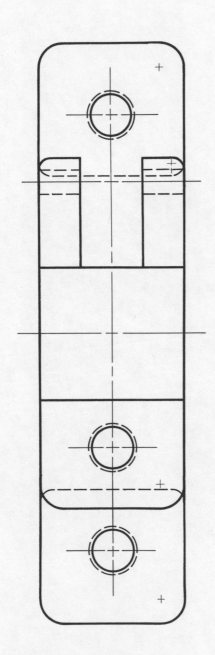
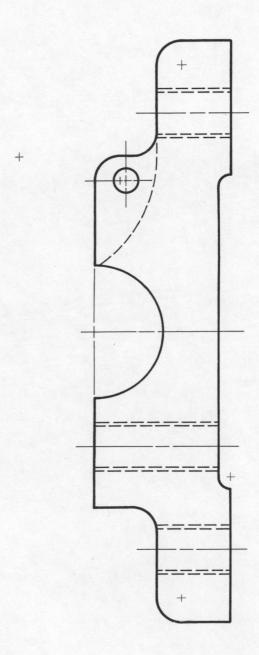

SCALE – FULL SIZE

2

GRADE	2	3	4	5	DIMENSIONING	DRAWN BY	FILE NO.	DRAWING
6	7	8	9	10	$\frac{1}{2}$	*Add Complete Dimensions*			Q-7

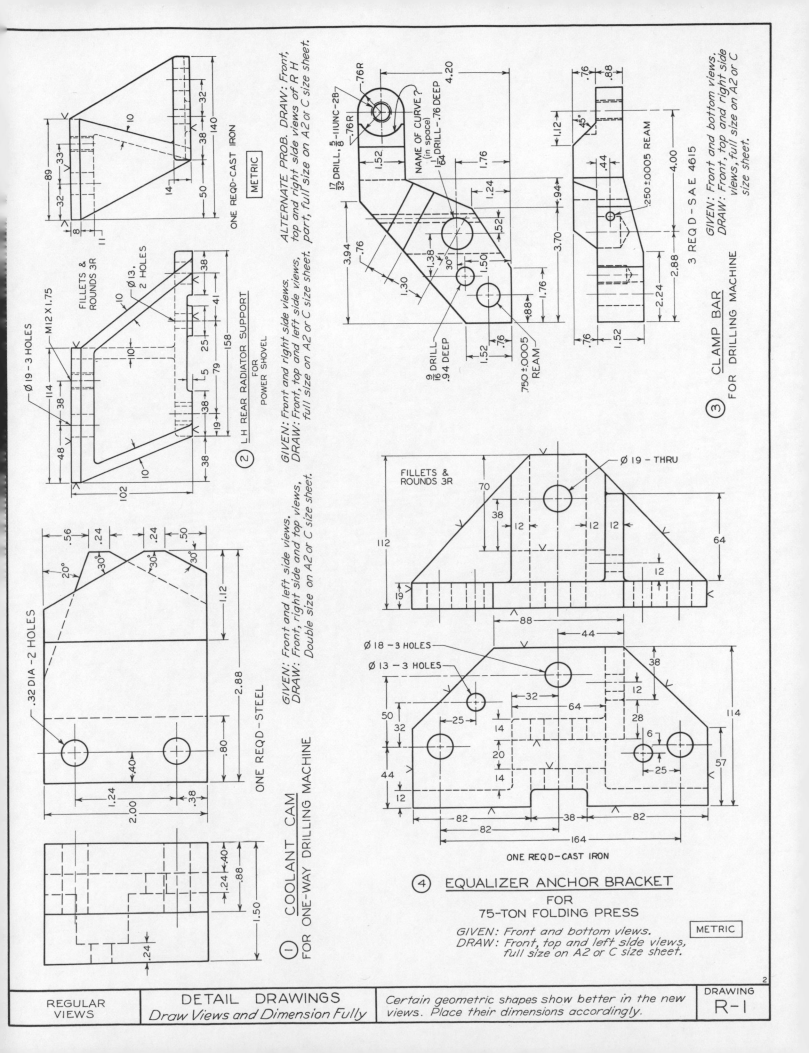

METRIC

ONE REQD-CAST IRON

89
33
32
8
11
32
140
38
32
10
14
50

FILLETS &
ROUNDS 3R

Ø 19 - 3 HOLES
M12×1.75
Ø 13,
2 HOLES
114
38
48
19
110
10
10
38
38
102
5
25
79
41
38
158

② LH REAR RADIATOR SUPPORT
FOR
POWER SHOVEL

ALTERNATE PROB. DRAW: Front,
top and right side views of R H
part, full size on A2 or C size sheet.

GIVEN: Front and right side views.
DRAW: Front, top and left side views,
full size on A2 or C size sheet.

.76R
17/32 DRILL, 5/8-11UNC-2B
.76R
1.52
NAME OF CURVE?
(in space)
1/64 DRILL -.76 DEEP
4.20
3.94
.76
1.30
1.38
30°
1.50
9/16 DRILL
.94 DEEP
1.52
.76
.750 ±.0005
REAM
.88
1.76
1.50
.52
1.24
1.76
3.70
.94

.76
.88
1.12
45°
.44
.250 ±.0005 REAM
4.00
3.70
.94
2.88
2.24
.76
1.52

3 REQD - SAE 4615

③ CLAMP BAR
FOR DRILLING MACHINE

GIVEN: Front and bottom views.
DRAW: Front, top and right side
views, full size on A2 or C
size sheet.

.56
.24
.24
.50
20°
30°
30°
30°
.32 DIA - 2 HOLES
1.12
2.88
.80
.40
1.24
.38
2.00

ONE REQD-STEEL

.24 .40
.88
.24
1.50
.24

① COOLANT CAM
FOR ONE-WAY DRILLING MACHINE

GIVEN: Front and left side views.
DRAW: Front, right side and top views,
Double size on A2 or C size sheet.

FILLETS &
ROUNDS 3R

Ø 19 - THRU
70
38
12
12
12
112
64
12
19
88

Ø 18 - 3 HOLES
Ø 13 - 3 HOLES
44
38
12
50
32
25
32
14
64
20
14
44
12
82
82
38
82
164
28
6
25
57
114

ONE REQD-CAST IRON

④ EQUALIZER ANCHOR BRACKET
FOR
75-TON FOLDING PRESS

GIVEN: Front and bottom views.
DRAW: Front, top and left side views,
full size on A2 or C size sheet.

METRIC

| REGULAR VIEWS | DETAIL DRAWINGS
Draw Views and Dimension Fully | *Certain geometric shapes show better in the new views. Place their dimensions accordingly.* | DRAWING
R-1 |

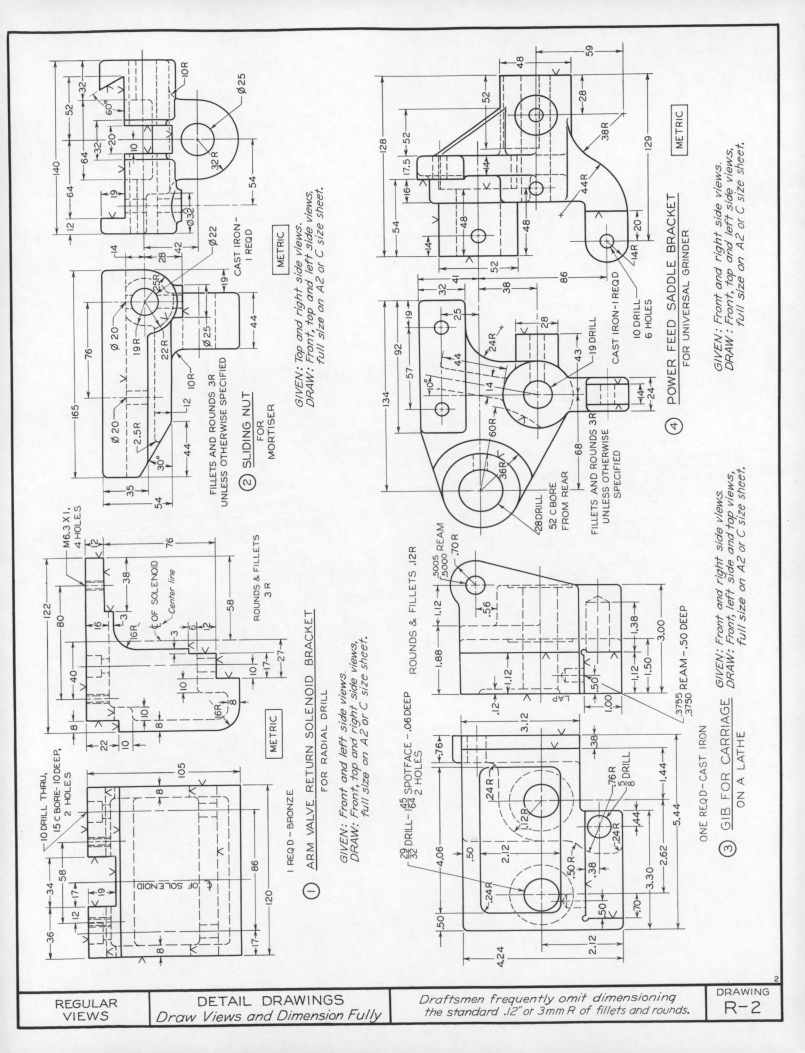

② SLIDING NUT
FOR MORTISER

FILLETS AND ROUNDS 3R
UNLESS OTHERWISE SPECIFIED

CAST IRON—I REQD

METRIC

GIVEN: Top and right side views.
DRAW: Front, top and left side views,
full size on A2 or C size sheet.

④ POWER FEED SADDLE BRACKET
FOR UNIVERSAL GRINDER

CAST IRON—I REQD

FILLETS AND ROUNDS 3R
UNLESS OTHERWISE SPECIFIED

10 DRILL
6 HOLES

METRIC

GIVEN: Front and right side views.
DRAW: Front, top and left side views,
full size on A2 or C size sheet.

① ARM VALVE RETURN SOLENOID BRACKET
FOR RADIAL DRILL

I REQD — BRONZE

ROUNDS & FILLETS 3R

METRIC

GIVEN: Front and left side views.
DRAW: Front, top and right side views,
full size on A2 or C size sheet.

③ GIB FOR CARRIAGE
ON A LATHE

ONE REQD—CAST IRON

ROUNDS & FILLETS .12R

GIVEN: Front and right side views.
DRAW: Front, left side and top views,
full size on A2 or C size sheet.

| REGULAR VIEWS | DETAIL DRAWINGS *Draw Views and Dimension Fully* | *Draftsmen frequently omit dimensioning the standard .12" or 3mm R of fillets and rounds.* | DRAWING R-2 |

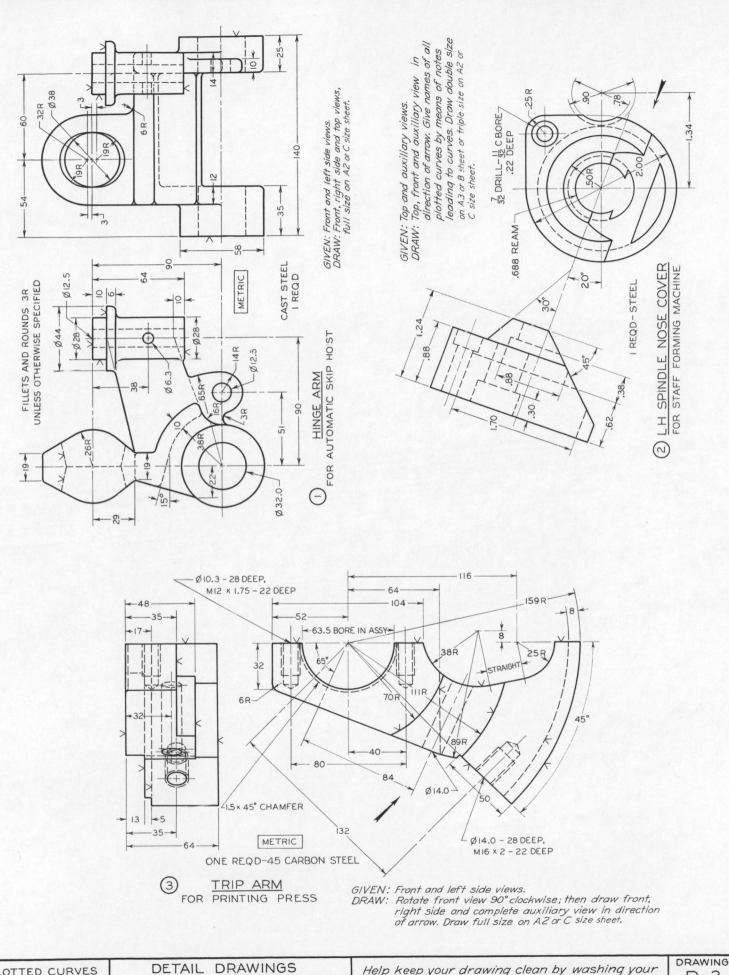

GIVEN: Front and left side views.
DRAW: Front, right side and top views, full size on A2 or C size sheet.

METRIC

CAST STEEL
1 REQ'D

FILLETS AND ROUNDS 3R
UNLESS OTHERWISE SPECIFIED

HINGE ARM
FOR AUTOMATIC SKIP HO'ST
①

GIVEN: Top and auxiliary views.
DRAW: Top, front and auxiliary view in direction of arrow. Give names of all plotted curves by means of notes leading to curves. Draw double size on A3 or B sheet or triple size on A2 or C size sheet.

LH SPINDLE NOSE COVER
FOR STAFF FORMING MACHINE
②
1 REQD–STEEL

½ DRILL–11/32 CBORE
.22 DEEP
.688 REAM

TRIP ARM
FOR PRINTING PRESS
③

METRIC

ONE REQD–45 CARBON STEEL

Ø10.3 – 28 DEEP,
M12 × 1.75 – 22 DEEP

63.5 BORE IN ASSY

1.5 × 45° CHAMFER

STRAIGHT

Ø14.0

Ø14.0 – 28 DEEP,
M16 × 2 – 22 DEEP

GIVEN: Front and left side views.
DRAW: Rotate front view 90° clockwise; then draw front, right side and complete auxiliary view in direction of arrow. Draw full size on A2 or C size sheet.

| PLOTTED CURVES AND AUX VIEWS | DETAIL DRAWINGS Draw Views and Dimension Fully | Help keep your drawing clean by washing your hands and removing your coat or sweater. | DRAWING R-3 |

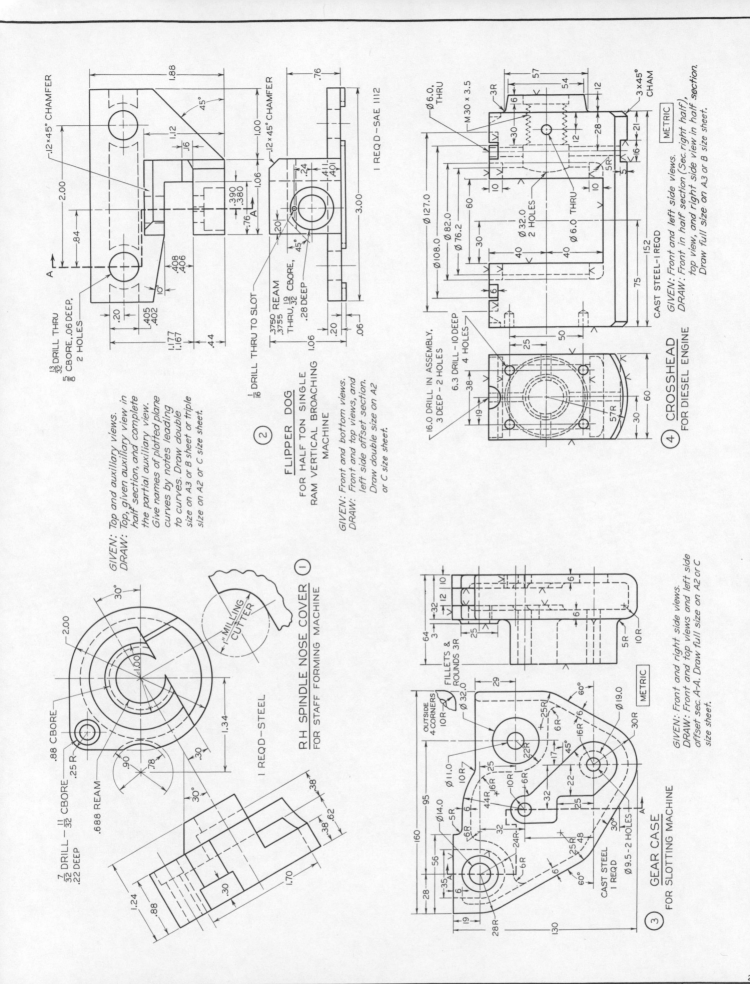

1.88

12×45° CHAMFER

45°

.12×45° CHAMFER

.76

1.12

.16

1.00

2.00

.84

.390
.380

1.06

.408
.406

.405
.402

1.177
1.167

.44

.20

.10

13/32 DRILL THRU
5/8 CBORE .06 DEEP,
2 HOLES

A

A

.24
.411
.401

.20
45°

.76

1.06

.20

.06

3.00

.3750 REAM
.3755 THRU, 19/32 CBORE,
.28 DEEP

1/16 DRILL THRU TO SLOT

I REQ'D – SAE 1112

GIVEN: Top and auxiliary views.
DRAW: Top, given auxiliary view in half section, and complete the partial auxiliary view. Give names of plotted plane curves by notes leading to curves. Draw double size on A3 or B sheet or triple size on A2 or C size sheet.

② **FLIPPER DOG**
FOR HALF TON SINGLE
RAM VERTICAL BROACHING
MACHINE

GIVEN: Front and bottom views.
DRAW: Front and top views, and left side offset section. Draw double size on A2 or C size sheet.

3R

Ø6.0, THRU

M30 × 3.5

6

57

54

12

3×45° CHAM

METRIC

30

12

28

Ø127.0

Ø108.0

Ø82.0

Ø76.2

Ø32.0
2 HOLES

Ø6.0 THRU

60

21

10

10

5R

5

16

40

40

30

152

75

50

25

16.0 DRILL IN ASSEMBLY,
3 DEEP – 2 HOLES

6.3 DRILL – 10 DEEP
4 HOLES

38

19

57R

30

60

CAST STEEL – I REQ'D

④ **CROSSHEAD**
FOR DIESEL ENGINE

GIVEN: Front and left side views.
DRAW: Front in half section (Sec. right half), top view, and right side view in half section. Draw full size on A3 or B size sheet.

30°

2.00

1.00

.88 CBORE

.25 R

.688 REAM

.90

.78

30°

30

7/32 DRILL – 11/32 CBORE
.22 DEEP

I REQ'D – STEEL

1" MILLING CUTTER

① **R H SPINDLE NOSE COVER**
FOR STAFF FORMING MACHINE

1.34

.38

.38

.62

1.24

.88

.30

1.70

30°

64

3

32

12

10

25

6

6

6

FILLETS & ROUNDS 3R

5R

10R

OUTSIDE 4 CORNERS
10R

Ø32.0

29

60°

60°

25R

6R

16R

76

Ø19.0

30R

Ø11.0

25

17

45°

22R

10R

6R

32

22

25

Ø14.0

44R

16R

25R
48

Ø9.5 – 2 HOLES

METRIC

95

160

56

28

35

6

6

24R

6R

32

6

60°

19

28R

130

A

CAST STEEL
I REQ'D

③ **GEAR CASE**
FOR SLOTTING MACHINE

GIVEN: Front and right side views.
DRAW: Front and top views and left side offset sec. A-A. Draw full size on A2 or C size sheet.

2

AUXILIARY VIEWS AND SECTIONS	DETAIL DRAWINGS *Draw Views and Dimension Fully*	*Sectional views may be dimensioned. Avoid placing dimensions within the hatched areas, if possible.*	DRAWING R-4

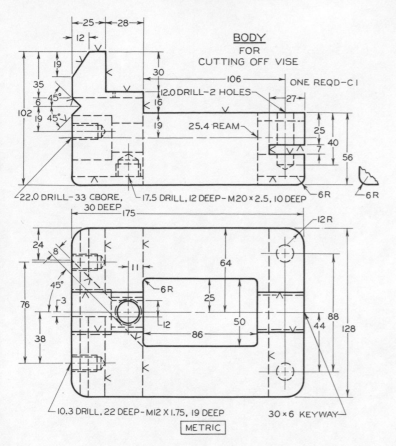

BODY
FOR
CUTTING OFF VISE

ONE REQD—C I

12.0 DRILL—2 HOLES

25.4 REAM

22.0 DRILL—33 CBORE, 30 DEEP
17.5 DRILL, 12 DEEP—M20 × 2.5, 10 DEEP
6R
6R

10.3 DRILL, 22 DEEP—M12 X 1.75, 19 DEEP
30 × 6 KEYWAY

12R
6R

METRIC

① GIVEN: Front and bottom views.
DRAW: Front full section, top and right side views.
Draw full size on A2 or C size sheet.

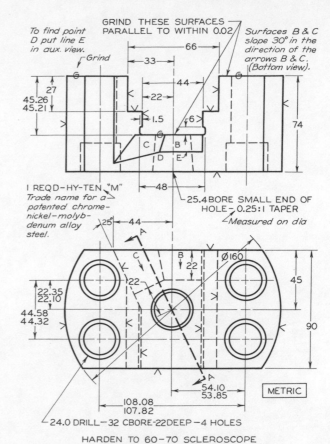

To find point D put line E in aux. view.
Grind

GRIND THESE SURFACES PARALLEL TO WITHIN 0.02

Surfaces B & C slope 30° in the direction of the arrows B & C. (Bottom view).

I REQD—HY-TEN "M"
Trade name for a patented chrome–nickel–molyb-denum alloy steel.

25.4 BORE SMALL END OF HOLE—0.25:1 TAPER
Measured on dia

24.0 DRILL—32 CBORE-22 DEEP—4 HOLES

METRIC

HARDEN TO 60–70 SCLEROSCOPE

The scleroscope is a hardness tester based on the rebound of a diamond-tipped hammer striking the metal being tested. The "60-70" is a reading on a special scale.

STRIPPER BLOCK
FOR FIXTURE ON 25-TON 2-COLUMN PRESS

② GIVEN: Front and bottom views.
DRAW: Front, top, auxiliary section A-A, and right side view in full section.
Draw full size on A2 or C size sheet.

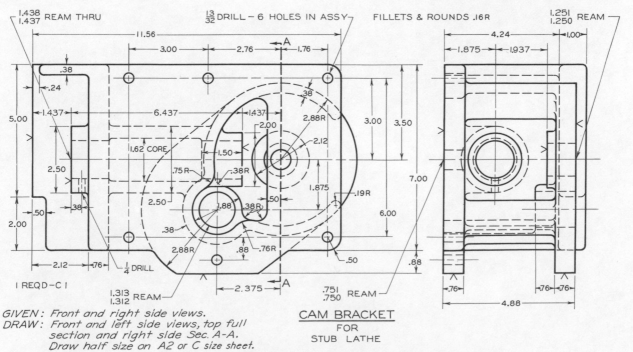

1.438 / 1.437 REAM THRU
13/32 DRILL – 6 HOLES IN ASSY
FILLETS & ROUNDS .16R
1.251 / 1.250 REAM

1.437
1.62 CORE
.75 R
.38 R
2.88R
1.88
.38
.38R
2.88R
.88 .76R
1/4 DRILL
.50
2.12 .76
1.313 / 1.312 REAM

.751 / .750 REAM

I REQD—C I

③ GIVEN: Front and right side views.
DRAW: Front and left side views, top full section and right side Sec. A-A.
Draw half size on A2 or C size sheet.

CAM BRACKET
FOR
STUB LATHE

SECTIONS | DETAIL DRAWINGS
Draw Views and Dimension Fully | Certain dimensions belong on the new view where certain geometric shapes will show best. | DRAWING R-5

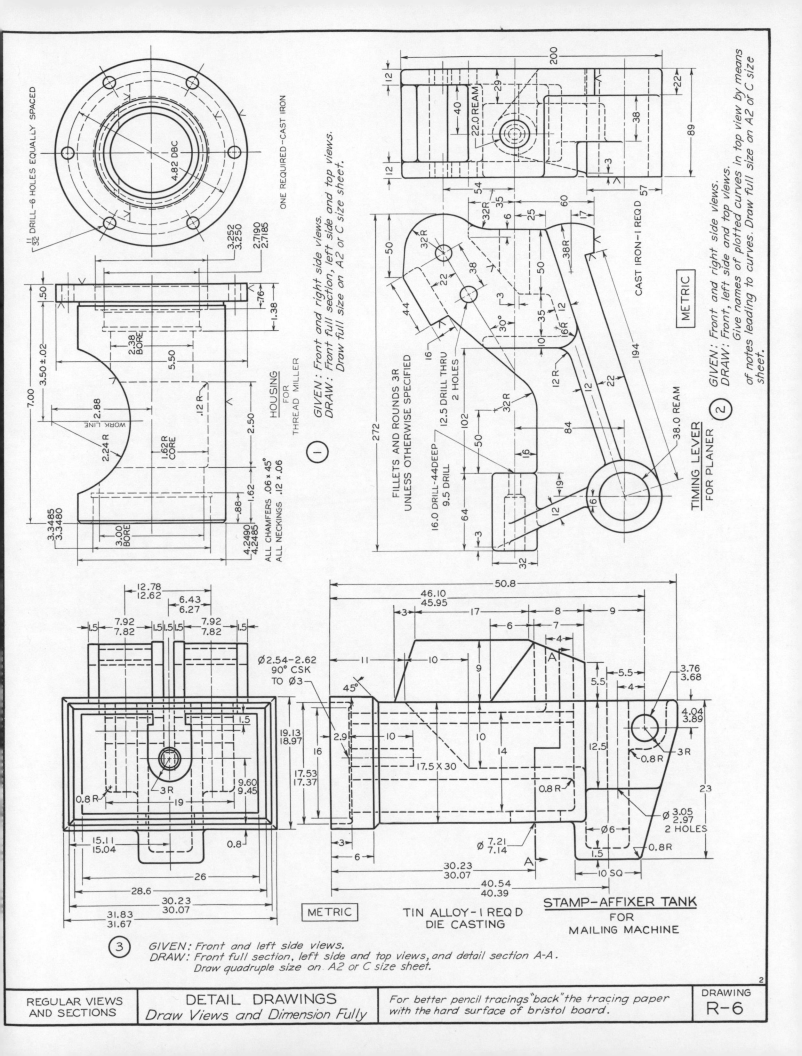

Drawing 1 — HOUSING FOR THREAD MILLER

ONE REQUIRED–CAST IRON

¹¹⁄₃₂ DRILL–6 HOLES EQUALLY SPACED

4.82 DBC

3.252 / 3.250
2.7190 / 2.7185

.50
.76
1.38

7.00
3.50 ± .02

2.38 BORE
5.50

2.88 WORK LINE
2.24 R
1.62 R CORE
.12 R

2.50
1.62
.88

3.3485 / 3.3480
3.00 BORE
4.2490 / 4.2485

ALL CHAMFERS .06 × 45°
ALL NECKINGS .12 × .06

① GIVEN: Front and right side views.
DRAW: Front full section, left side and top views.
Draw full size on A2 or C size sheet.

Drawing 2 — TIMING LEVER FOR PLANER

CAST IRON–1 REQ'D   METRIC

200
12
12
29
40
22.0 REAM
22
89
38
3
54
57

32R
35
6
25
60
17
50
32R
22
38
38R
44
3
50
30°
35
194
16
10
12R
12
16
6R
12.5 DRILL THRU 2 HOLES
102
32R
12
22
16.0 DRILL–44 DEEP 9.5 DRILL
50
84
38.0 REAM
64
16
19
12
6
3
32

FILLETS AND ROUNDS 3R
UNLESS OTHERWISE SPECIFIED

272

② GIVEN: Front and right side views.
DRAW: Front, left side and top views.
Give names of plotted curves in top view by means
of notes leading to curves. Draw full size on A2 or C size
sheet.

Drawing 3 — STAMP–AFFIXER TANK FOR MAILING MACHINE

TIN ALLOY–1 REQ'D DIE CASTING   METRIC

12.78 / 12.62
6.43 / 6.27
7.92 / 7.82
1.5 1.5 1.5
7.92 / 7.82
1.5

Ø2.54–2.62 90° CSK TO Ø3

1.5
19.13 / 18.97
16
17.53 / 17.37

9.60 / 9.45
3R
19
0.8 R

15.11 / 15.04
0.8
26
28.6
30.23 / 30.07
31.83 / 31.67

50.8
46.10 / 45.95
3
17
8
9
6
7
4
11
10
9
A'
5.5
5.5
4
45°
2.9
10
10
14
12.5
17.5 × 30
0.8 R
0.8 R
Ø 7.21 / 7.14
3
6
A
30.23 / 30.07
40.54 / 40.39
10 SQ

3.76 / 3.68
4.04 / 3.89
3R
Ø 3.05 / 2.97 2 HOLES
Ø6
1.5
0.8R
23

③ GIVEN: Front and left side views.
DRAW: Front full section, left side and top views, and detail section A-A.
Draw quadruple size on A2 or C size sheet.

REGULAR VIEWS AND SECTIONS | DETAIL DRAWINGS *Draw Views and Dimension Fully* | *For better pencil tracings "back" the tracing paper with the hard surface of bristol board.* | DRAWING R-6

2

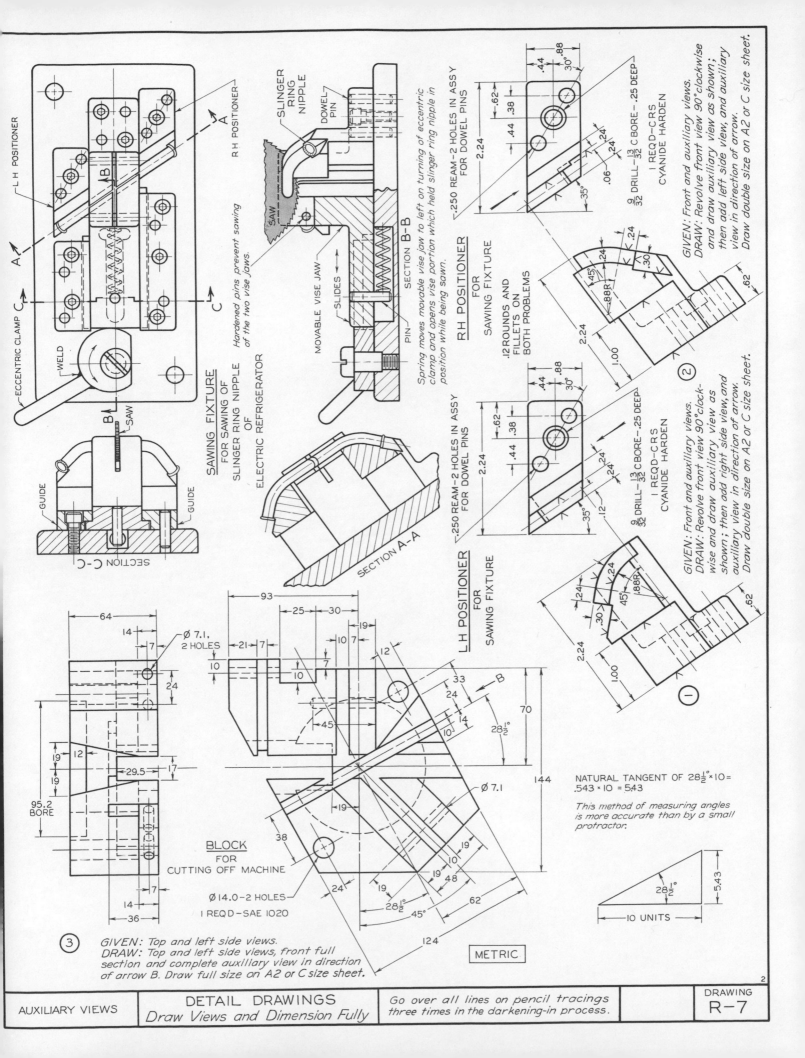

LH POSITIONER

RH POSITIONER

A

ECCENTRIC CLAMP C

WELD

B

SAW

GUIDE

GUIDE

SECTION C-C

SAWING FIXTURE
FOR SAWING OF
SLINGER RING NIPPLE
OF
ELECTRIC REFRIGERATOR

SLINGER
RING NIPPLE

DOWEL
PIN

SAW

MOVABLE VISE JAW

SLIDES

PIN

SECTION B-B

*Hardened pins prevent sawing
of the two vise jaws.*

*Spring moves movable vise jaw to left on turning of eccentric
clamp and opens vise portion which held slinger ring nipple in
position while being sawn.*

SECTION A-A

.250 REAM-2 HOLES IN ASSY
FOR DOWEL PINS

.88
.44
30°
.62
.44 .38
2.24
.24
.24
.06
9/32 DRILL-13/32 C BORE-.25 DEEP
1 REQD-CRS
CYANIDE HARDEN
35°

GIVEN: *Front and auxiliary views.*
DRAW: *Revolve front view 90° clockwise
and draw auxiliary view as shown;
then add left side view, and auxiliary
view in direction of arrow.
Draw double size on A2 or C size sheet.*

.24
.24
.30
45°
.88R
2.24
1.00
.62
.12 ROUNDS AND
FILLETS ON
BOTH PROBLEMS

②

RH POSITIONER
FOR
SAWING FIXTURE

.250 REAM-2 HOLES IN ASSY
FOR DOWEL PINS

.88
.44
30°
.62
.44 .38
2.24
.24
.24
9/32 DRILL-13/32 CBORE-.25 DEEP
1 REQD-CRS
CYANIDE HARDEN
.12
35°

GIVEN: *Front and auxiliary views.*
DRAW: *Revolve front view 90° clockwise and draw auxiliary view as
shown; then add right side view, and
auxiliary view in direction of arrow.
Draw double size on A2 or C size sheet.*

.24
.24
.30
45°
.88R
2.24
1.00
.62

①

LH POSITIONER
FOR
SAWING FIXTURE

64
14
7
Ø 7.1,
2 HOLES
24
19
12
19
29.5
17
95.2
BORE
7
14
36

93
25
30
21
7
19
10 7
10
10
7
45
12
33
24
14
10
B
28½°
70
Ø 7.1
Ø 14.0-2 HOLES
1 REQD-SAE 1020
38
19
19
24
19
10 19
48
28½°
45°
62
144
124

BLOCK
FOR
CUTTING OFF MACHINE

NATURAL TANGENT OF 28½°×10=
.543 × 10 = 5.43

*This method of measuring angles
is more accurate than by a small
protractor.*

5.43
28½°
10 UNITS

③ GIVEN: *Top and left side views.*
DRAW: *Top and left side views, front full
section and complete auxiliary view in direction
of arrow B. Draw full size on A2 or C size sheet.*

METRIC

| AUXILIARY VIEWS | DETAIL DRAWINGS
Draw Views and Dimension Fully | *Go over all lines on pencil tracings
three times in the darkening-in process.* | DRAWING
R-7 |

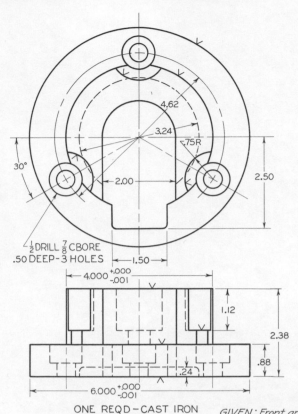

4.62
3.24
.75R
2.00
2.50
30°

½ DRILL ⅞ CBORE
.50 DEEP-3 HOLES
1.50

4.000 +.000 -.001
1.12
2.38
.88
.24
6.000 +.000 -.001

ONE REQD—CAST IRON

① **CENTERING BUSHING**
FOR A KEY SEATER

GIVEN: Front and bottom views.
DRAW: Front and left side views, and right side full section. Draw full size on A3 or B size sheet.

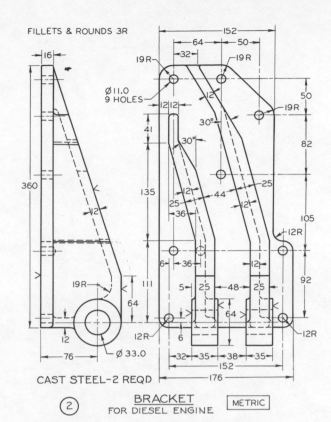

FILLETS & ROUNDS 3R

152
64
50
16
19R
32
19R
Ø11.0
9 HOLES
12 12
12
19R
50
41
30°
30°
360
135
12
44
25
25
36
12
105
12R
19R
6
36
12
92
111
5
25
48
25
64
12
6
Ø33.0
76
32 35 38 35
152
176

CAST STEEL-2 REQD

② **BRACKET**
FOR DIESEL ENGINE `METRIC`

GIVEN: Front and left side views.
DRAW: Revolve front view 90° clockwise and let it be a top view ; then add front and right side views. Draw half size on A3 or B size sheet.

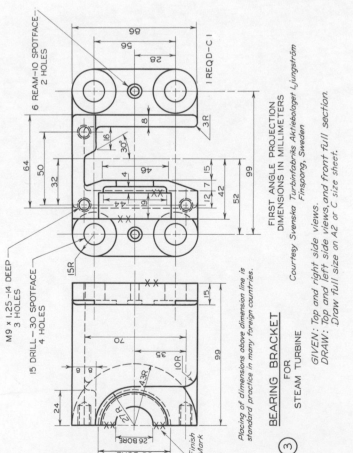

6 REAM-10 SPOTFACE
2 HOLES

86
56
28
I REQD-C I
8
3R
16
30°
64
50
4
46
32
12 7 15
42
99
44
52
19
M9 x 1.25-14 DEEP
3 HOLES
15 DRILL-30 SPOTFACE
4 HOLES
15R
15
70
35
10R
99
24
4.3R
17R
26 BORE
50 BORE
Finish Mark

FIRST ANGLE PROJECTION
DIMENSIONS IN MILLIMETERS

Courtesy Svenska Turbinfabriks Aktiebolaget Ljungström Finspong, Sweden

③ **BEARING BRACKET**
FOR STEAM TURBINE

GIVEN: Top and right side views.
DRAW: Top and left side views, and front full section. Draw full size on A2 or C size sheet.

Placing of dimensions above dimension line is standard practice in many foreign countries.

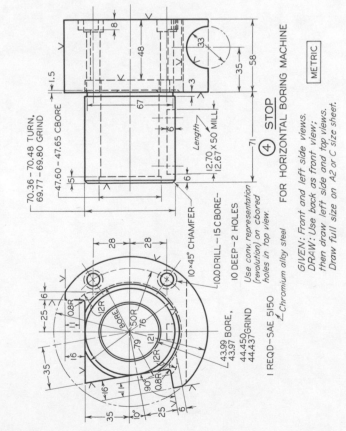

8
33
48
35
58
1.5
3
70.36 - 70.48 TURN,
69.77 - 69.80 GRIND
47.60 - 47.65 CBORE
67
6
71
Length
12.70 x 50 MILL
12.67 x 50 MILL
10 x 45° CHAMFER
10.0 DRILL-15 C BORE-
10 DEEP-2 HOLES
Use conv. representation (revolution) on cbored holes in top view.

28 28
25 6
0.8R
12R
BORE
50R
79 76 121
12R
43.99 BORE,
43.97 BORE,
44.450 GRIND
44.437 GRIND
25 6
6
16
35
90°
0.8R
35
16
11
25 6
10°

I REQD-SAE 5150
Chromium alloy steel

④ **STOP**
FOR HORIZONTAL BORING MACHINE `METRIC`

GIVEN: Front and left side views.
DRAW: Use back as front view; then draw left side and top views. Draw full size on A2 or C size sheet.

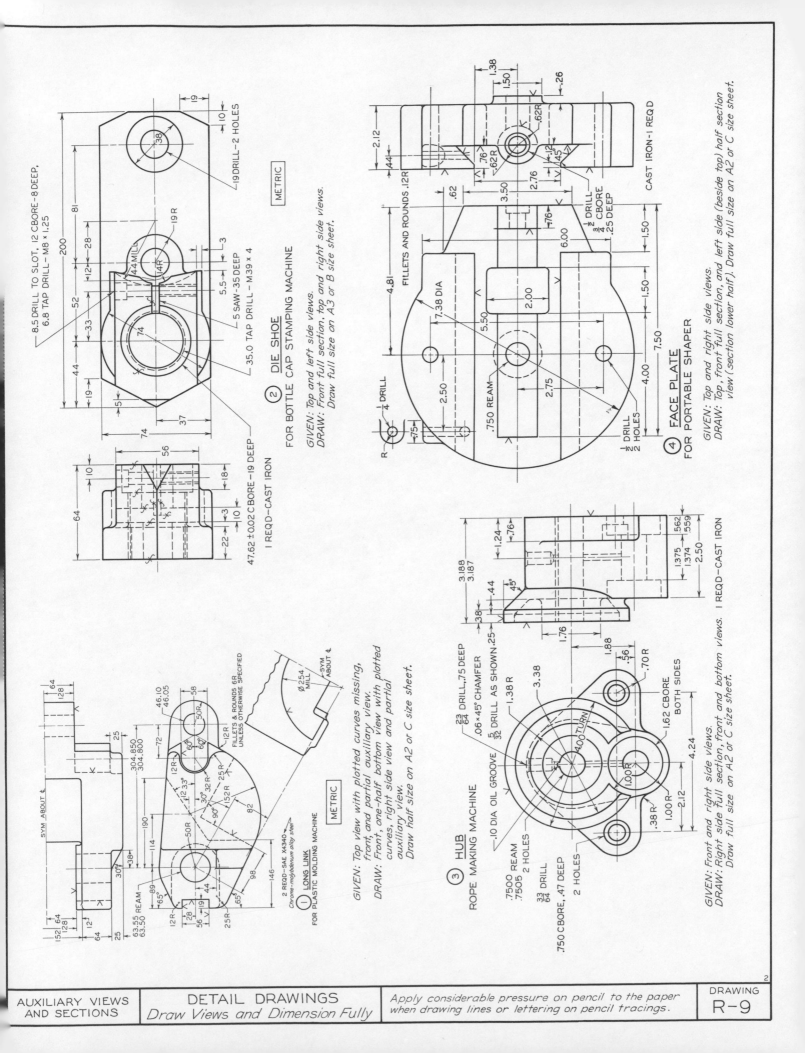

① LONG LINK
FOR PLASTIC MOLDING MACHINE

METRIC

GIVEN: Top view with plotted curves missing, front and partial auxiliary view.
DRAW: Front, one-half bottom view with plotted curves, right side view and partial auxiliary view.
Draw half size on A2 or C size sheet.

2 REQD–SAE X4340
Chrome–molybdenum alloy steel

FILLETS & ROUNDS 6R
UNLESS OTHERWISE SPECIFIED

Ø 254
MILL

SYM ABOUT ₵

② DIE SHOE
FOR BOTTLE CAP STAMPING MACHINE

METRIC

GIVEN: Top and left side views.
DRAW: Front full section, top and right side views.
Draw full size on A3 or B size sheet.

1 REQD–CAST IRON

8.5 DRILL TO SLOT, 12 CBORE–8 DEEP,
6.8 TAP DRILL–M8 × 1.25

19 DRILL–2 HOLES

19 R

14R

44 MILL

5 SAW–35 DEEP

35.0 TAP DRILL–M39 × 4

47.62 ±0.02 CBORE–19 DEEP

③ HUB
ROPE MAKING MACHINE

GIVEN: Front and right side views.
DRAW: Right side full section, front and bottom views. 1 REQD–CAST IRON
Draw full size on A2 or C size sheet.

.10 DIA OIL GROOVE

23/64 DRILL, .75 DEEP
.06×45° CHAMFER

5/32 DRILL AS SHOWN

4.00 TURN

1.62 CBORE
BOTH SIDES

.7500 REAM
.7505   2 HOLES

33/64 DRILL

.750 CBORE, .47 DEEP
2 HOLES

④ FACE PLATE
FOR PORTABLE SHAPER

GIVEN: Top and right side views.
DRAW: Top, front full section, and left side (beside top) half section view (section lower half). Draw full size on A2 or C size sheet.

CAST IRON–1 REQD

FILLETS AND ROUNDS .12R

1/4 DRILL
3/4 CBORE
.25 DEEP

7.38 DIA

.750 REAM

1/4 DRILL

1/2 DRILL
2 HOLES

1 CENTERING BRACKET

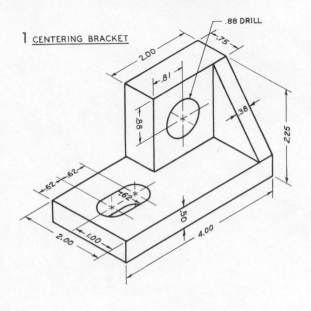

.88 DRILL
2.00
.75
.81
.88
.38
2.25
.62 .62
.62
.50
2.00 1.00
4.00

2 HINGE   METRIC

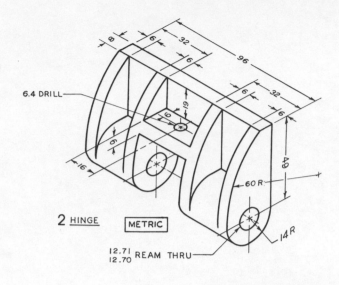

8 6 32
6
96
6.4 DRILL
6
19
6
32
6
16
49
60 R
14 R
12.71 REAM THRU
12.70

3 GRIPPING JAW

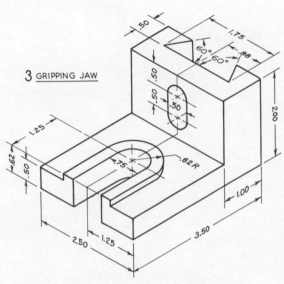

.50
1.75
60° 60°
.88
.50
.50
.50
2.00
1.25
.62 R
.62
.50
.75
1.00
2.50 1.25
3.50

4 GUIDE BLOCK

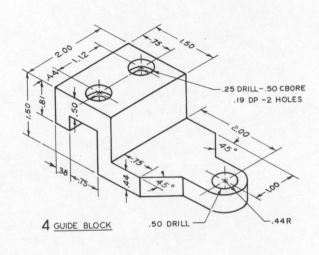

2.00
1.12
.75 1.50
.44
.81
.50
1.50
.25 DRILL-.50 CBORE
.19 DP -2 HOLES
2.00
4.5°
.38
.75
.75
.44
4.5°
1.00
.50 DRILL
.44 R

5 HINGE BASE   METRIC

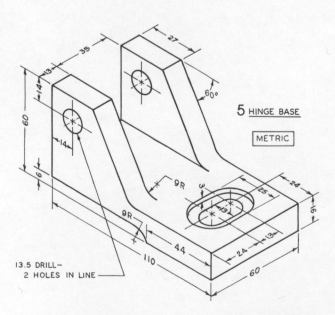

35
27
13
14
60°
60
14
9 R
3
25
24
6
9 R
13
16
13.5 DRILL-
2 HOLES IN LINE
44
24
110
60

6 PIVOT PLATE   METRIC

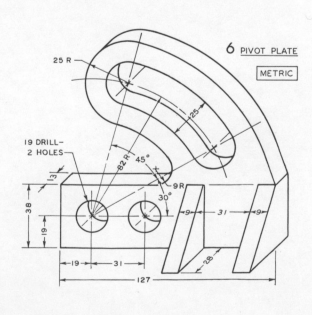

25 R
25
19 DRILL-
2 HOLES
82 R
45°
13
9 R
30°
38
9
31
9
19
19
31
28
127

DETAIL DRAWINGS

DRAW OR SKETCH THE NECESSARY VIEWS OF THE OBJECT
ASSIGNED. DIMENSION COMPLETELY

DRAWING
R-10

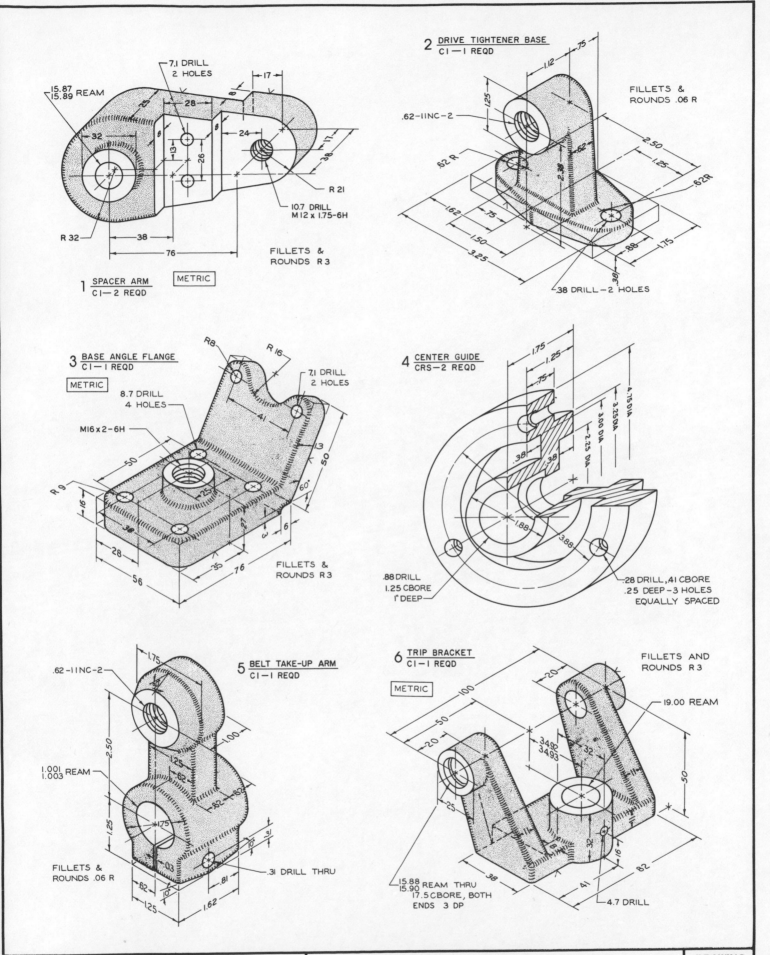

15.87 REAM
15.89

7.1 DRILL
2 HOLES

28

17

32

23

28

8

13

26

24

8

R 21

10.7 DRILL
M12 x 1.75-6H

R 32

38

76

1 SPACER ARM
C1—2 REQD

METRIC

FILLETS &
ROUNDS R 3

---

2 DRIVE TIGHTENER BASE
C1—1 REQD

1.12

.75

1.25

FILLETS &
ROUNDS .06 R

.62-11NC-2

.62

.62 R

2.35

2.50

1.25

.62R

1.62

.75

.75

1.50

.88

1.75

3.25

.38

.38 DRILL—2 HOLES

---

3 BASE ANGLE FLANGE
C1—1 REQD

METRIC

R8

R 16

7.1 DRILL
2 HOLES

8.7 DRILL
4 HOLES

41

M16 x 2-6H

.13

50

50

60°

.25

R 9

16

61

6

38

3

28

35

76

56

FILLETS &
ROUNDS R 3

---

4 CENTER GUIDE
CRS—2 REQD

1.75

1.25

.75

4.75 DIA

3.25 DIA

3.00 DIA

.38

.38

2.25 DIA

1.88

3.88

.88 DRILL
1.25 CBORE
1" DEEP

28 DRILL, .41 CBORE
.25 DEEP—3 HOLES
EQUALLY SPACED

---

5 BELT TAKE-UP ARM
C1—1 REQD

1.75

.62-11NC-2

1.00

.25

1.001 REAM
1.003

2.50

.62

.82

.85

.75

1.25

.03

.31

.10

FILLETS &
ROUNDS .06 R

.62

.10

.81

1.25

1.62

.31 DRILL THRU

---

6 TRIP BRACKET
C1—1 REQD

METRIC

FILLETS AND
ROUNDS R 3

20

100

50

19.00 REAM

20

34.92
34.93

32

50

25

11

32

16

38

41

82

15.88 REAM THRU
15.90
17.5 CBORE, BOTH
ENDS 3 DP

4.7 DRILL

DETAIL DRAWINGS

DRAW OR SKETCH THE NECESSARY VIEWS OF THE OBJECT
ASSIGNED. DIMENSION COMPLETELY

DRAWING

R-11

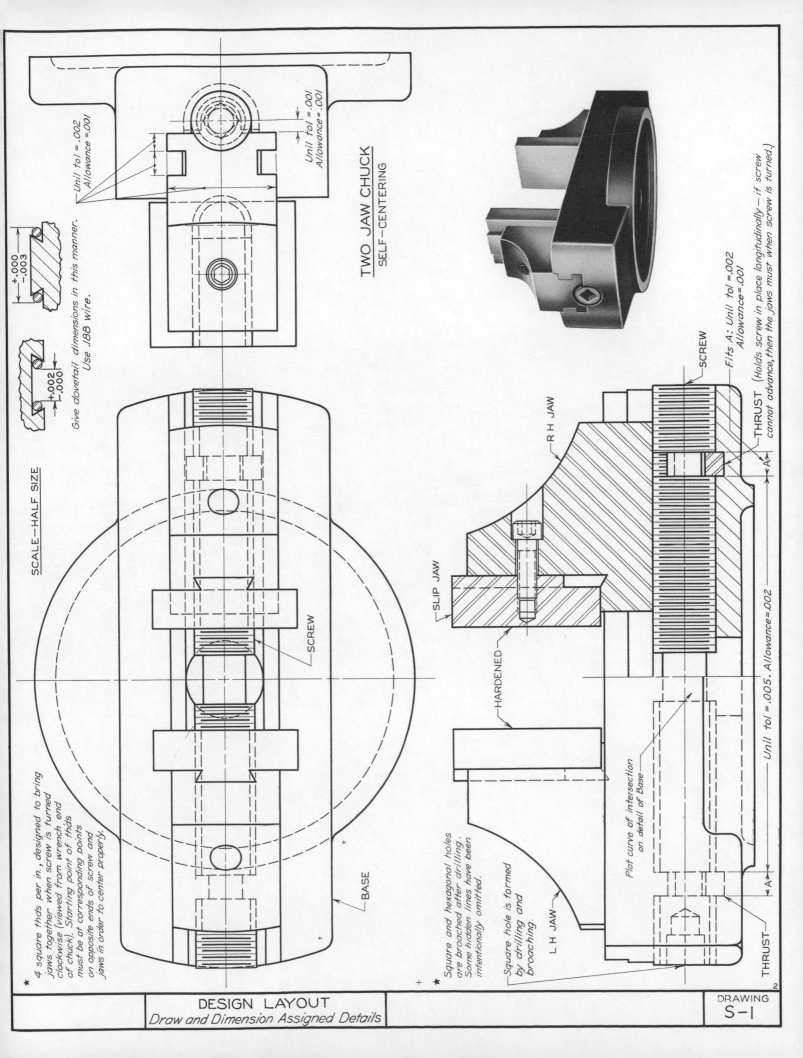

SCALE—HALF SIZE

Unit tol.=.002
Allowance=.001

+.000
-.003

+.002
-.000

Give dovetail dimensions in this manner.
Use .188 wire.

Unit tol.=.001
Allowance=.001

TWO JAW CHUCK
SELF—CENTERING

★ 4 square thds per in., designed to bring
jaws together when screw is turned
clockwise (viewed from wrench end
of chuck). Starting point of thds
must be at corresponding points
on opposite ends of screw and
jaws in order to center properly.

SCREW

BASE

★ Square and hexagonal holes
are broached after drilling.
Some hidden lines have been
intentionally omitted.

Square hole is formed
by drilling and
broaching.

L H JAW

R H JAW

SLIP JAW

HARDENED

SCREW

Fits A: Unit tol.=.002
Allowance=.001

THRUST (Holds screw in place longitudinally — if screw
cannot advance, then the jaws must when screw is turned.)

A

A

Plot curve of intersection
on detail of Base

Unit tol.=.005. Allowance=.002

THRUST

DESIGN LAYOUT
Draw and Dimension Assigned Details

DRAWING
S—1

2

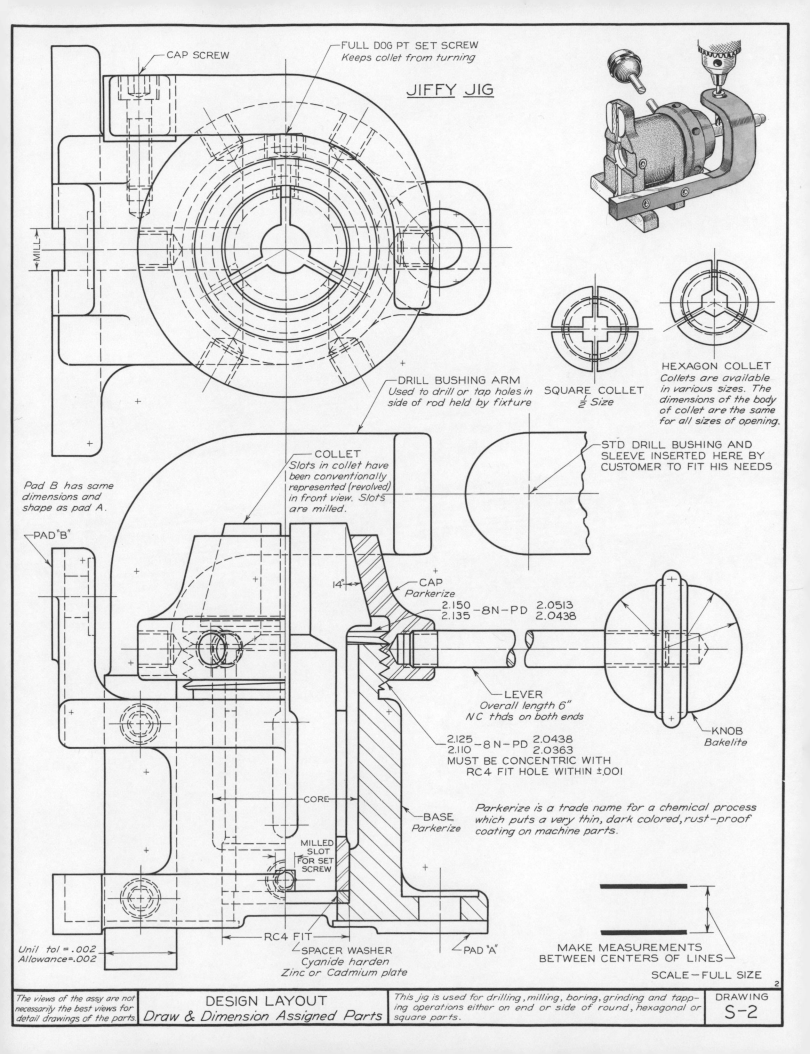

CAP SCREW

FULL DOG PT SET SCREW
Keeps collet from turning

JIFFY JIG

MILL

SQUARE COLLET
½ Size

HEXAGON COLLET
Collets are available
in various sizes. The
dimensions of the body
of collet are the same
for all sizes of opening.

DRILL BUSHING ARM
Used to drill or tap holes in
side of rod held by fixture

STD DRILL BUSHING AND
SLEEVE INSERTED HERE BY
CUSTOMER TO FIT HIS NEEDS

Pad B has same
dimensions and
shape as pad A.

PAD "B"

COLLET
Slots in collet have
been conventionally
represented (revolved)
in front view. Slots
are milled.

14°

CAP
Parkerize

$\frac{2.150}{2.135}$ —8N—PD $\frac{2.0513}{2.0438}$

LEVER
Overall length 6"
NC thds on both ends

KNOB
Bakelite

$\frac{2.125}{2.110}$ —8N—PD $\frac{2.0438}{2.0363}$
MUST BE CONCENTRIC WITH
RC4 FIT HOLE WITHIN ±.001

CORE

BASE
Parkerize

Parkerize is a trade name for a chemical process
which puts a very thin, dark colored, rust-proof
coating on machine parts.

MILLED
SLOT
FOR SET
SCREW

Unit tol =.002
Allowance=.002

RC4 FIT

SPACER WASHER
Cyanide harden
Zinc or Cadmium plate

PAD "A"

MAKE MEASUREMENTS
BETWEEN CENTERS OF LINES

SCALE—FULL SIZE

The views of the assy are not
necessarily the best views for
detail drawings of the parts.

DESIGN LAYOUT
Draw & Dimension Assigned Parts

This jig is used for drilling, milling, boring, grinding and tapping operations either on end or side of round, hexagonal or
square parts.

DRAWING
S-2

2

I.

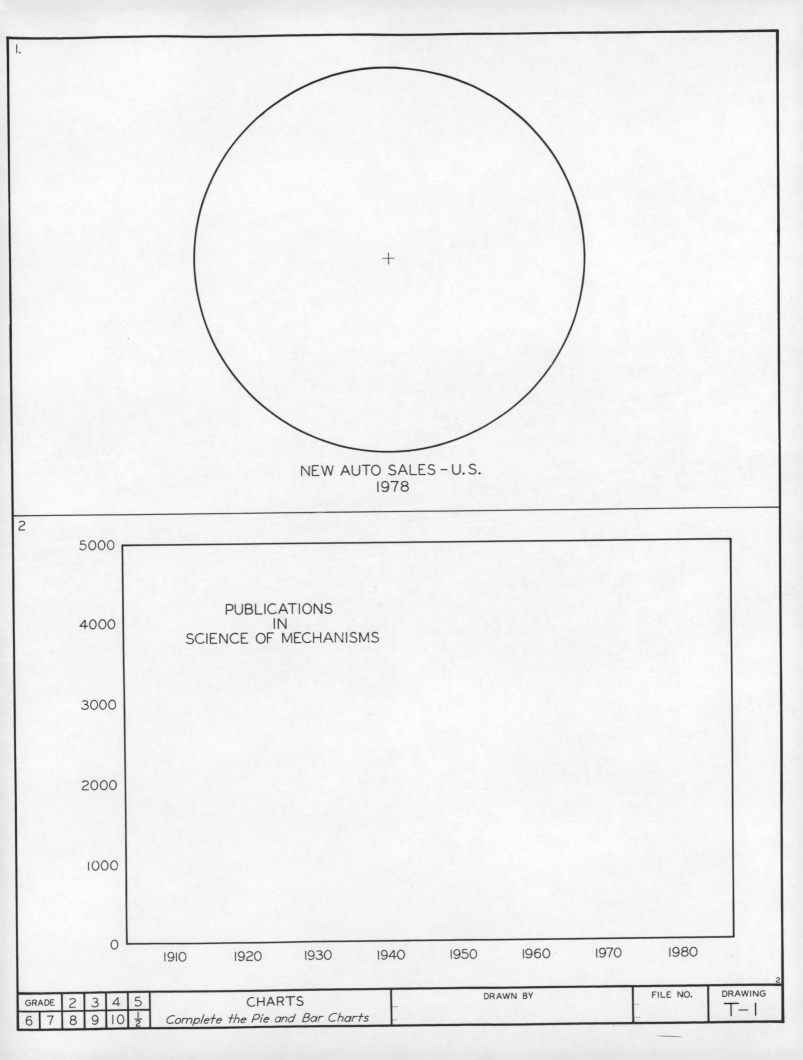

NEW AUTO SALES – U.S.
1978

2

PUBLICATIONS
IN
SCIENCE OF MECHANISMS

5000

4000

3000

2000

1000

0

1910   1920   1930   1940   1950   1960   1970   1980

GRADE	2	3	4	5	CHARTS	DRAWN BY	FILE NO.	DRAWING
6	7	8	9	10	½	Complete the Pie and Bar Charts			T–I

I.

EXPECTED LIFE OF GENERAL PURPOSE SNAP-ACTION SWITCH

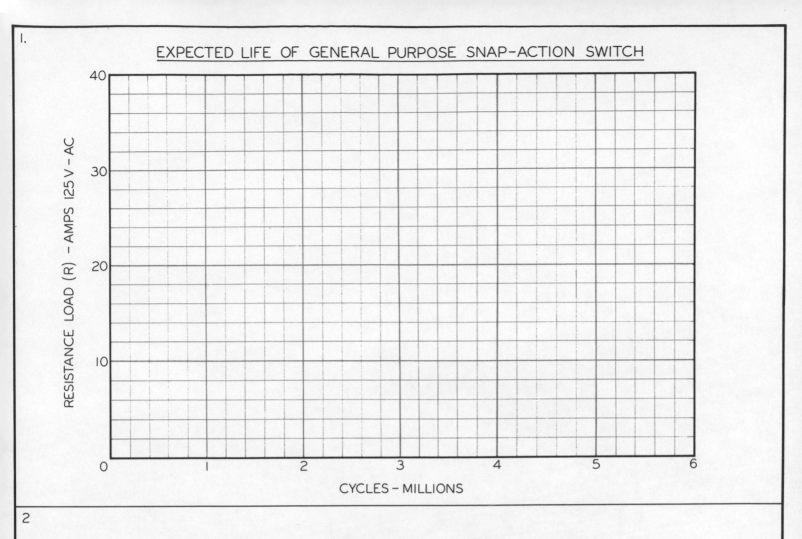

RESISTANCE LOAD (R) – AMPS 125 V – AC

CYCLES – MILLIONS

2

EXPECTED LIFE OF GENERAL PURPOSE SNAP-ACTION SWITCH

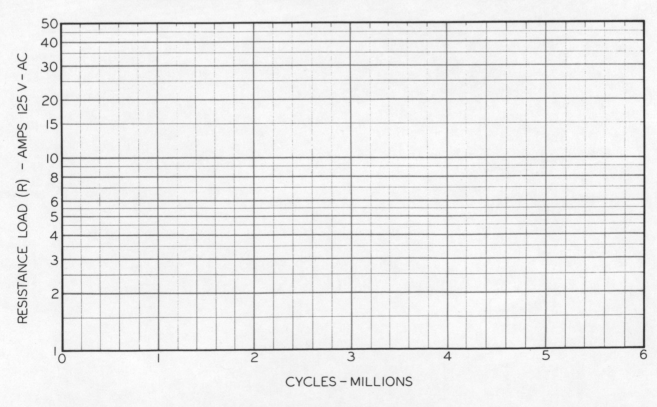

RESISTANCE LOAD (R) – AMPS 125 V – AC

CYCLES – MILLIONS

GRADE	2	3	4	5	ENGINEERING GRAPHS	DRAWN BY	FILE NO.	DRAWING
6	7	8	9	10	½	*Draw the Rect. & Semi-log Coord. Graphs.*			T-2

I.

18 —
16 —
14 —
12 —
10 —
8 —
6 —
4 —
2 —
0 —

X

$X + y + 2 = Z$

2

10 —
9 —
8 —
7 —
6 —
5 —
4 —
3 —
2 —
1 —
0 —

AMPERES (I)

$V = IR$

GRADE	2	3	4	5	NOMOGRAPHS	DRAWN BY	FILE NO.	DRAWING
6	7	8	9	10	$\frac{1}{2}$	Complete the Parallel Scale & N-Charts.			U-I

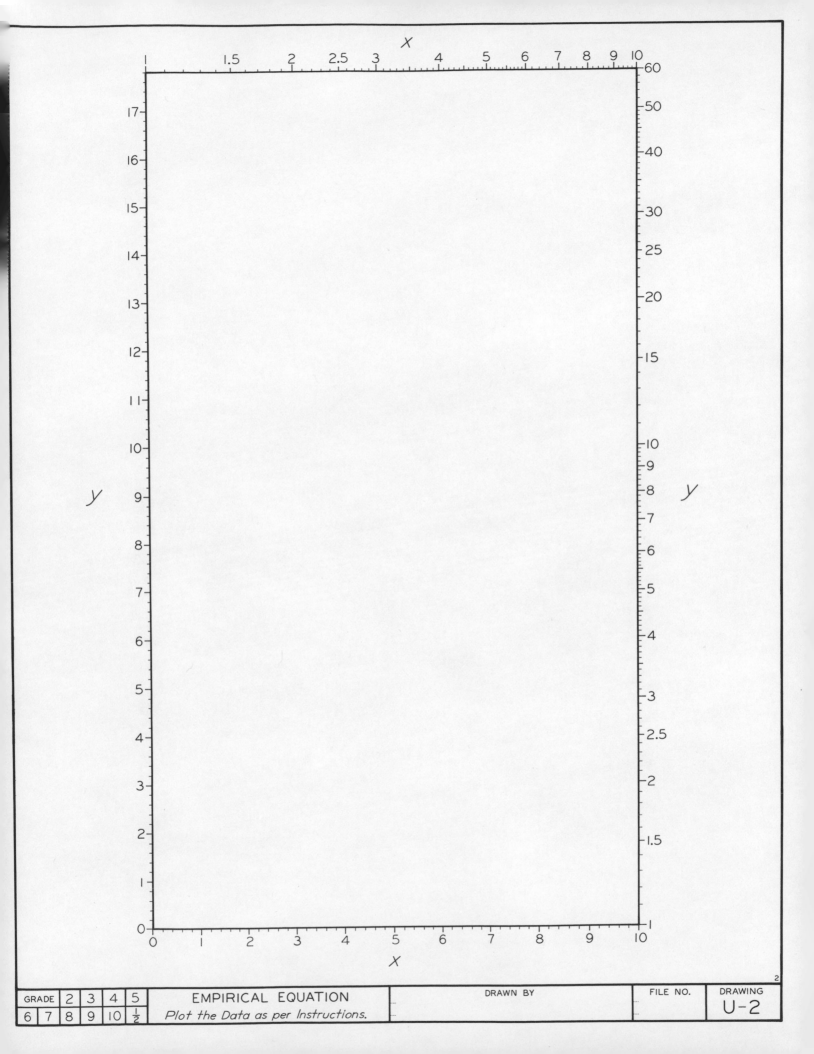

X

17 —
16 —
15 —
14 —
13 —
12 —
11 —
10 —
y   9 —
8 —
7 —
6 —
5 —
4 —
3 —
2 —
1 —
0 —

0   1   2   3   4   5   6   7   8   9   10

X

1.5   2   2.5   3   4   5   6   7   8   9   10

60
50
40
30
25
20
15
10
9
8
7
6
5
4
3
2.5
2
1.5
1

y

GRADE	2	3	4	5	EMPIRICAL EQUATION	DRAWN BY	FILE NO.	DRAWING
6	7	8	9	10	$\frac{1}{2}$	Plot the Data as per Instructions.			U-2

$2y = x^2 + 2x - 16$
$x - 2y - 4 = 0$

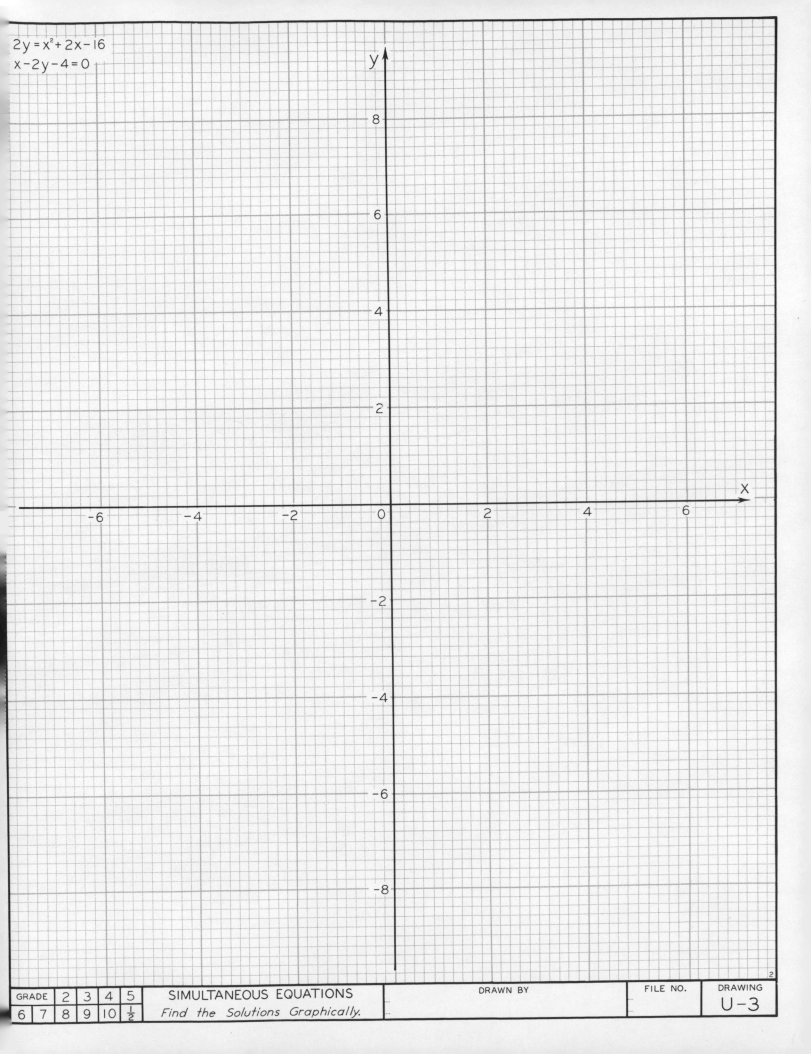

	y
	8
	6
	4
	2

-6   -4   -2   0   2   4   6   x

	-2
	-4
	-6
	-8

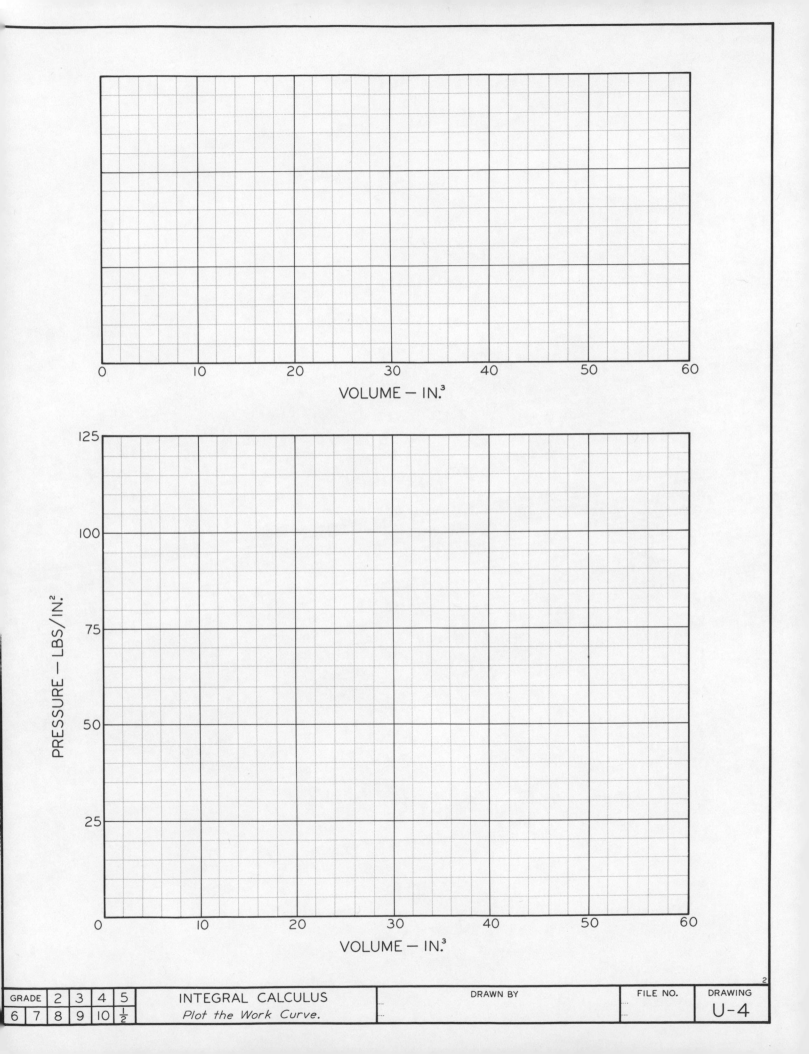

VOLUME — IN.^{3}

PRESSURE — LBS/IN.^{2}

VOLUME — IN.^{3}

GRADE	2	3	4	5	INTEGRAL CALCULUS	DRAWN BY	FILE NO.	DRAWING
6	7	8	9	10	$\frac{1}{2}$	Plot the Work Curve.			U-4

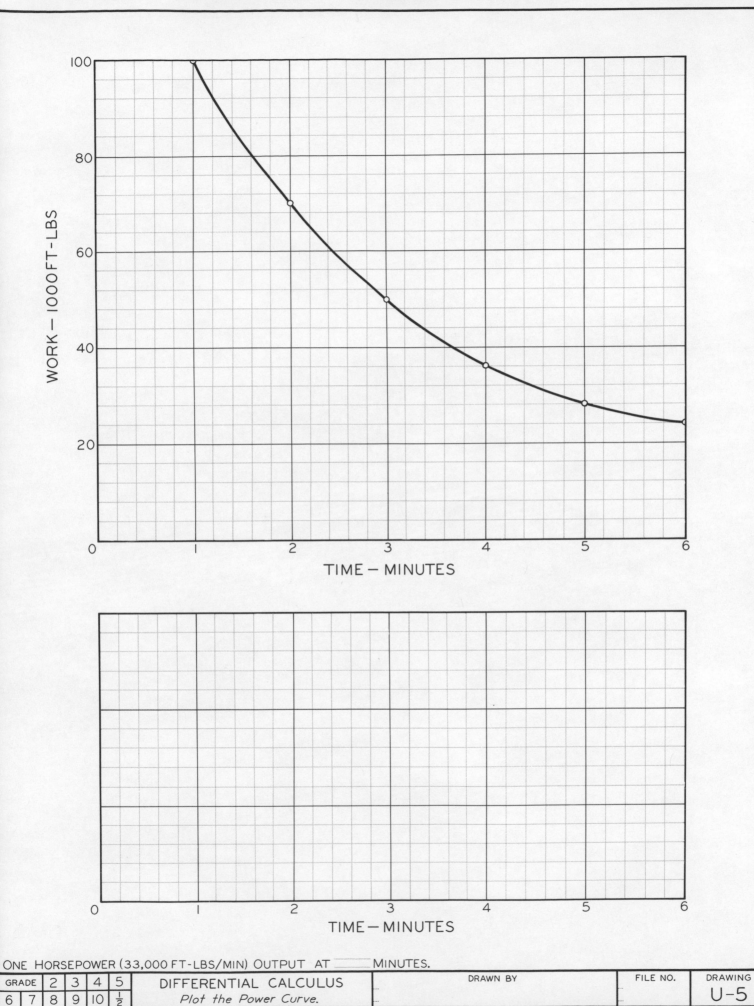

WORK — 1000 FT-LBS

TIME — MINUTES

TIME — MINUTES

ONE HORSEPOWER (33,000 FT-LBS/MIN) OUTPUT AT _____ MINUTES.

GRADE	2	3	4	5	DIFFERENTIAL CALCULUS	DRAWN BY	FILE NO.	DRAWING
6	7	8	9	10	½	Plot the Power Curve.			U-5

Complete the table of TERMS by entering the letter identifiers of the matching descriptions.

TERMS

CURSOR	
DIGITIZER TABLET	
GRAPHIC PRIMITIVE	
PIXEL	
RESOLUTION	
RASTER DISPLAY	
RAM	
DEBUG	
HARD COPY	
ANALOG	
DIGITAL	
CAE	
COMMAND	
PLOTTER	
HARD DISK	
VECTOR	
COORDINATE SYSTEM	
MENU	
BIT	
MOUSE	
SOFTWARE	
JOY STICK	
WINDOW	
TRANSFORM	
BYTE	
LIGHTPEN	

Descriptions

A Handheld pointing device for pick and coordinate entry

B Computer program to perform specific tasks

C Counts in discrete steps or digits

D Smallest unit of digital information

E Collection of commands for selection

F Device to convert analog picture to coordinate digital data

G Fundamental drawing entity

H Picture element dot in a display grid

I Random Access Memory - volatile physical memory

J Continuous measurements without steps

K Computer assisted engineering

L Group of 8 bits commonly used to represent a character

M Paper printout

N Hand controlled lever used as input device

O Smallest spacing between CRT display elements

P Convert an image into a proper display format

Q Directed line segment with magnitude

R Flicker-free scanned CRT surface

S A bounded rectangular area on screen

T A visual tracking symbol

U Handheld photosensitive input device

V Control signal

W Correct errors

X Non-volatile external storage device

Y Hard copy device for vector drawing

Z Common reference system for spatial relationships

GRADE	2	3	4	5	COMPUTER-AIDED DRAFTING	DRAWN BY	FILE NO.	DRAWING
	6	7	8	9	10	$\frac{1}{2}$	TERMS AND DESCRIPTIONS			V—1

① Complete the table by defining X and Y coordinates of the given points.

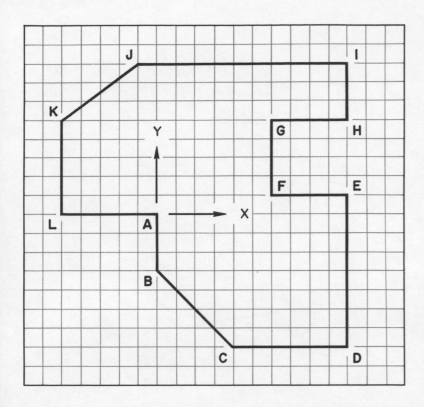

Point	Coordinate	
	X	Y
A		
B		
C		
D		
E		
F		
G		
H		
I		
J		
K		
L		

② Plot the given points on the grid and draw the view.

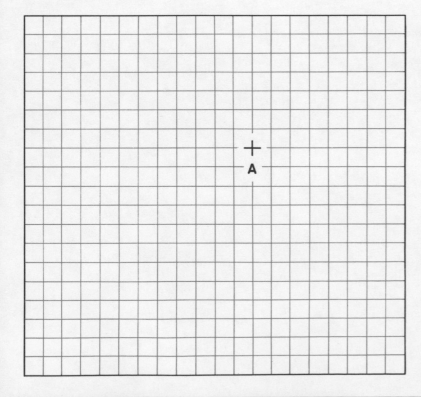

Point	Coordinate	
	X	Y
A	0	0
B	0	4
C	− 5	4
D	− 5	− 2
E	−10	− 6
F	−10	−10
G	− 2	−10
H	6	− 4
I	6	3
J	3	3
K	3	− 2
L	0	− 4

GRADE	2	3	4	5	COMPUTER-AIDED DRAFTING	DRAWN BY	FILE NO.	DRAWING
	6	7	8	9	10	½	TWO-DIMENSIONAL COORDINATE PLOT			V—2

① Complete the table for drawing the object.

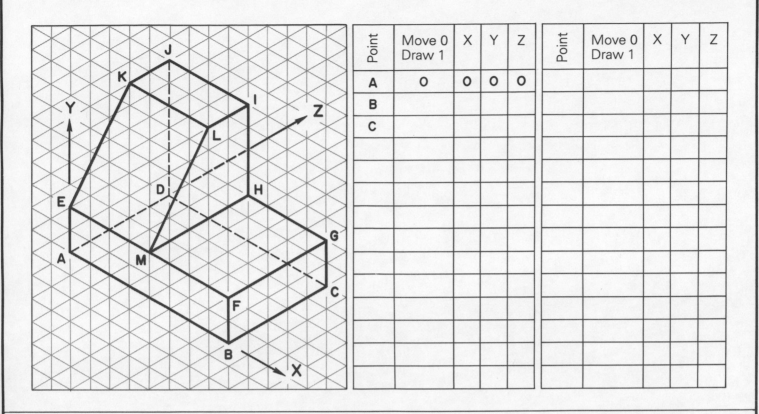

Point	Move 0 Draw 1	X	Y	Z	Point	Move 0 Draw 1	X	Y	Z
A	0	0	0	0					
B									
C									

② Draw the object based on the data given in the table.

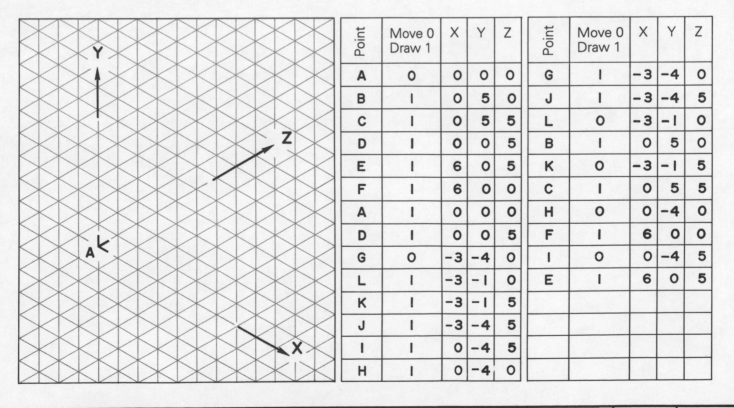

Point	Move 0 Draw 1	X	Y	Z	Point	Move 0 Draw 1	X	Y	Z
A	0	0	0	0	G	1	-3	-4	0
B	1	0	5	0	J	1	-3	-4	5
C	1	0	5	5	L	0	-3	-1	0
D	1	0	0	5	B	1	0	5	0
E	1	6	0	5	K	0	-3	-1	5
F	1	6	0	0	C	1	0	5	5
A	1	0	0	0	H	0	0	-4	0
D	1	0	0	5	F	1	6	0	0
G	0	-3	-4	0	I	0	0	-4	5
L	1	-3	-1	0	E	1	6	0	5
K	1	-3	-1	5					
J	1	-3	-4	5					
I	1	0	-4	5					
H	1	0	-4	0					

Complete the table by entering the Menu Selections used for generating the drawing.

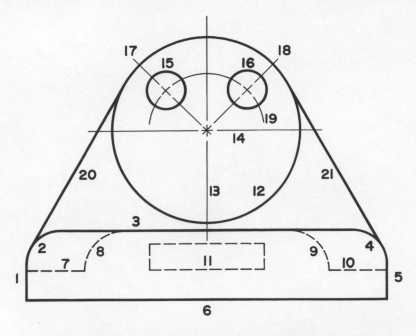

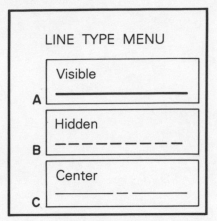

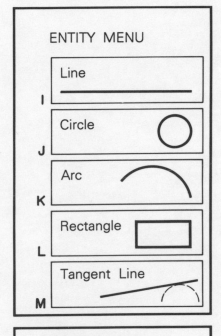

Entity	Line type menu selection	Entity menu selection	Construction menu selection
1			
2			
3			
4			
5			
6			
7			
8			
9			
10			
11			
12			
13			
14			
15			
16			
17			
18			
19			
20			
21			

GRADE	2	3	4	5	COMPUTER-AIDED DRAFTING	DRAWN BY	FILE NO.	DRAWING
6	7	8	9	10	$\frac{1}{2}$	MENU USAGE			V—4

VIEW COORDINATES are the coordinate values of the object as assigned with respect to the computer screen, with X , Y and Z axes positioned as shown below . The coordinates remain the same irrespective of the view selected on the screen .

Axis	Position	Positive Direction
X	Horizontal	To the right
Y	Vertical	Toward the top
Z	Perpendicular to the screen	Outward from the screen

Description of WORLD COORDINATES

WORLD COORDINATES are the coordinate values of the object as assigned with respect to the axes of the object . The X , Y and Z axes are positioned as shown, such that for the top view the X axis is horizontal to the right, the Y axis is vertical to the top and the Z axis is perpendicular to the screen positioned outwards . The coordinates in relation to the screen change according to the view selected on the screen.

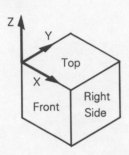

Complete the tables by entering the VIEW and WORLD COORDINATES of the given points of the object .

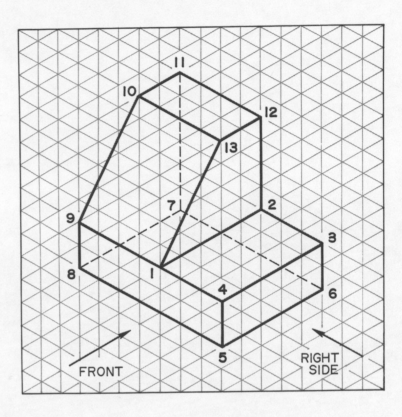

FRONT VIEW

Points	View Coordinates			World Coordinates		
	X	Y	Z	X	Y	Z
1						
2						
3						
4						
5						
6						
7						
8						
9						
10						
11						
12						
13						

RIGHT SIDE VIEW

Points	View Coordinates			World Coordinates		
	X	Y	Z	X	Y	Z
1						
2						
3						
4						
5						
6						
7						
8						
9						
10						
11						
12						
13						

GRADE	2	3	4	5	COMPUTER-AIDED DRAFTING	DRAWN BY	FILE NO.	DRAWING
6	7	8	9	10	$\frac{1}{2}$	COORDINATE SYSTEMS			V—5

DRAWING

DO NOT SCALE DRAWING

BREAK ALL SHARP CORNERS

GRADE	4	5	6	PART	DRAWN BY	TRACED BY	APPROVED	DATE	SCALE	FILE NO.	DRAWING
7	8	9	10	½							

DRAWING

BREAK ALL SHARP CORNERS

	DRAWING
FILE NO.	
SCALE	
DATE	
APPROVED	
TRACED BY	

DRAWN BY

DO NOT SCALE DRAWING

GRADE	4	5	6	PART
7	8	9	10	
			½	

DRAWING

DO NOT SCALE DRAWING

BREAK ALL SHARP CORNERS

GRADE	4	5	6	PART	DRAWN BY	TRACED BY	APPROVED	DATE	SCALE	FILE NO.	DRAWING
7	8	9	10	½							

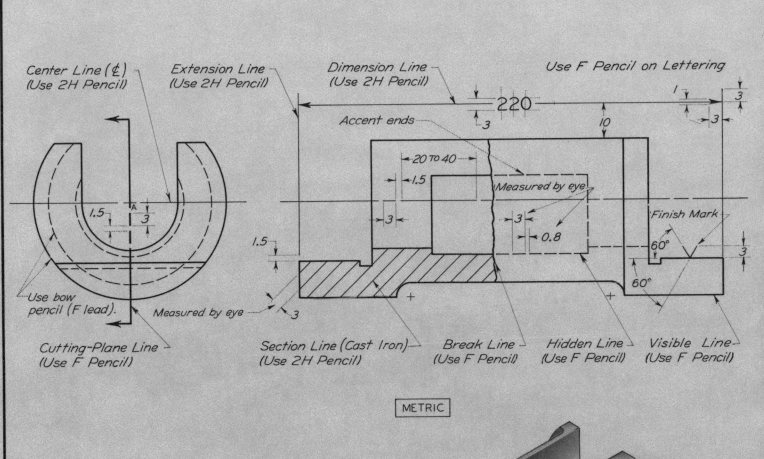

Center Line (℄)
(Use 2H Pencil)

Extension Line
(Use 2H Pencil)

Dimension Line
(Use 2H Pencil)

Use F Pencil on Lettering

Accent ends

220

20 TO 40

1.5

Measured by eye

1.5

3

0.8

Finish Mark

60°

60°

10

3

3

3

1.5

3

Use bow
pencil (F lead).

Measured by eye

Cutting-Plane Line
(Use F Pencil)

Section Line (Cast Iron)
(Use 2H Pencil)

Break Line
(Use F Pencil)

Hidden Line
(Use F Pencil)

Visible Line
(Use F Pencil)

METRIC

TAPE SEGMENT
FOR
GEAR GRINDER

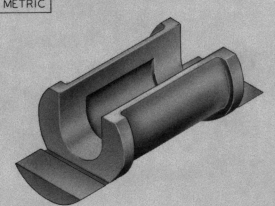

A

NOTE:
All pencil lines except construction and projection lines are dark lines. Copy the given views
starting at A. Omit all inclined lettering. Make your dark lines as black as those above.

GRADE	2	3	4	5	ALPHABET OF LINES		DRAWN BY		FILE NO.		DRAWING
	6	7	8	9	10	½	*Draw Given Views*				A-4

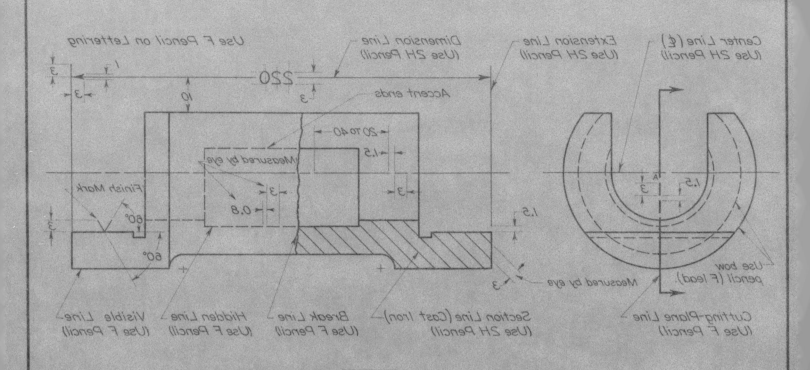

METRIC

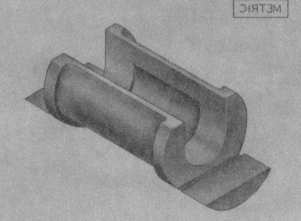

TAPE SEGMENT
FOR
GEAR GRINDER

NOTE:
All pencil lines except construction and projection lines are dark lines. Copy the given views
starting at A. Omit all inclined lettering. Make your dark lines as black as those above.

GRADE	5	4	3	2	ALPHABET OF LINES	DRAWN BY	FILE NO.	DRAWING
6	10	9	8	7	Draw Given Views			A-4

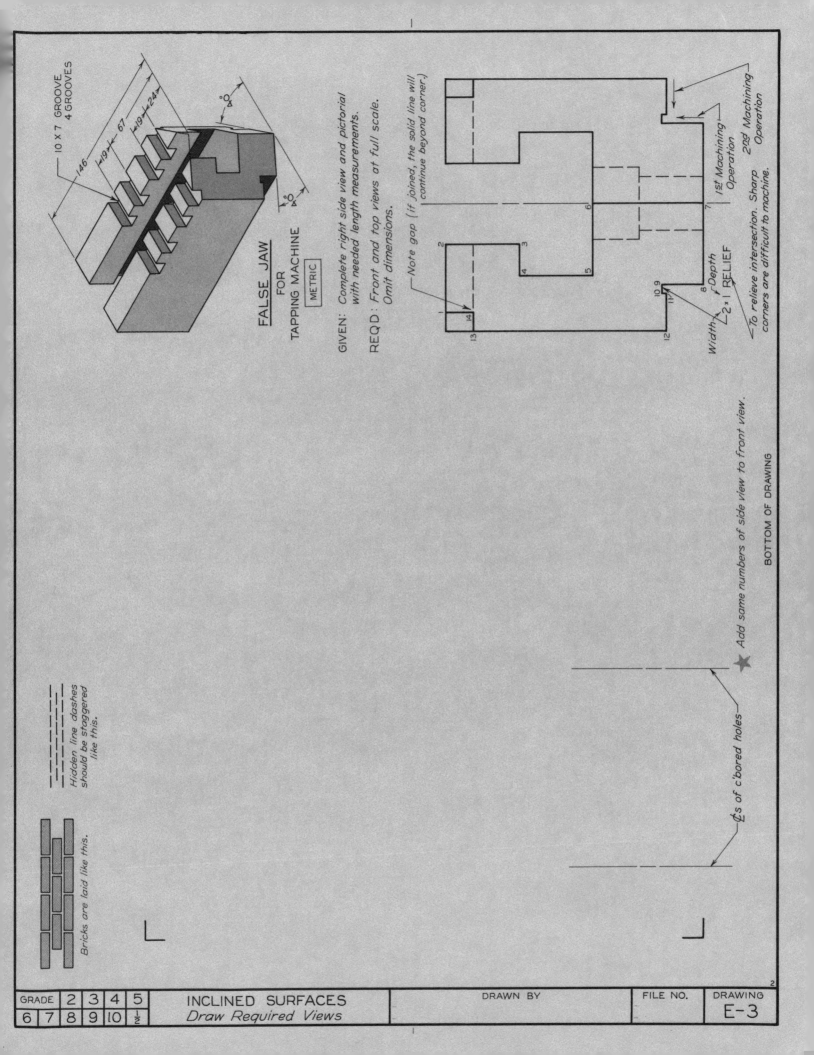

10 X 7 GROOVE
4 GROOVES

146
19 67 19 24

0°

0°

FALSE JAW
FOR
TAPPING MACHINE
METRIC

GIVEN: *Complete right side view and pictorial with needed length measurements.*

REQD: *Front and top views at full scale. Omit dimensions.*

Note gap (if joined, the solid line will continue beyond corner.)

1st Machining Operation

2nd Machining Operation

To relieve intersection. Sharp corners are difficult to machine.

2 × I RELIEF

Width
Depth

1
2
3
4
5
6
7
8
9
10
11
12
13
14

BOTTOM OF DRAWING

Add same numbers of side view to front view.

Cs of c'bored holes

Hidden line dashes should be staggered like this.

Bricks are laid like this.

GRADE	2	3	4	5
	6	7	8	9	10	½

INCLINED SURFACES
Draw Required Views

DRAWN BY

FILE NO.

DRAWING
E-3

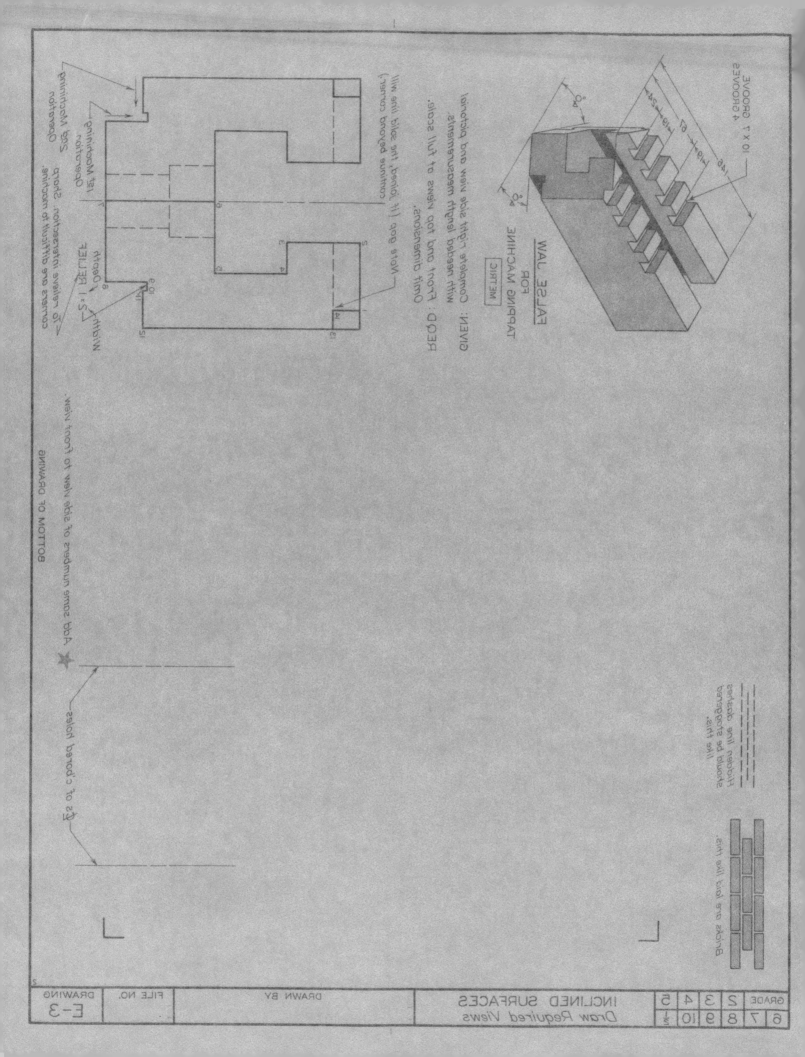

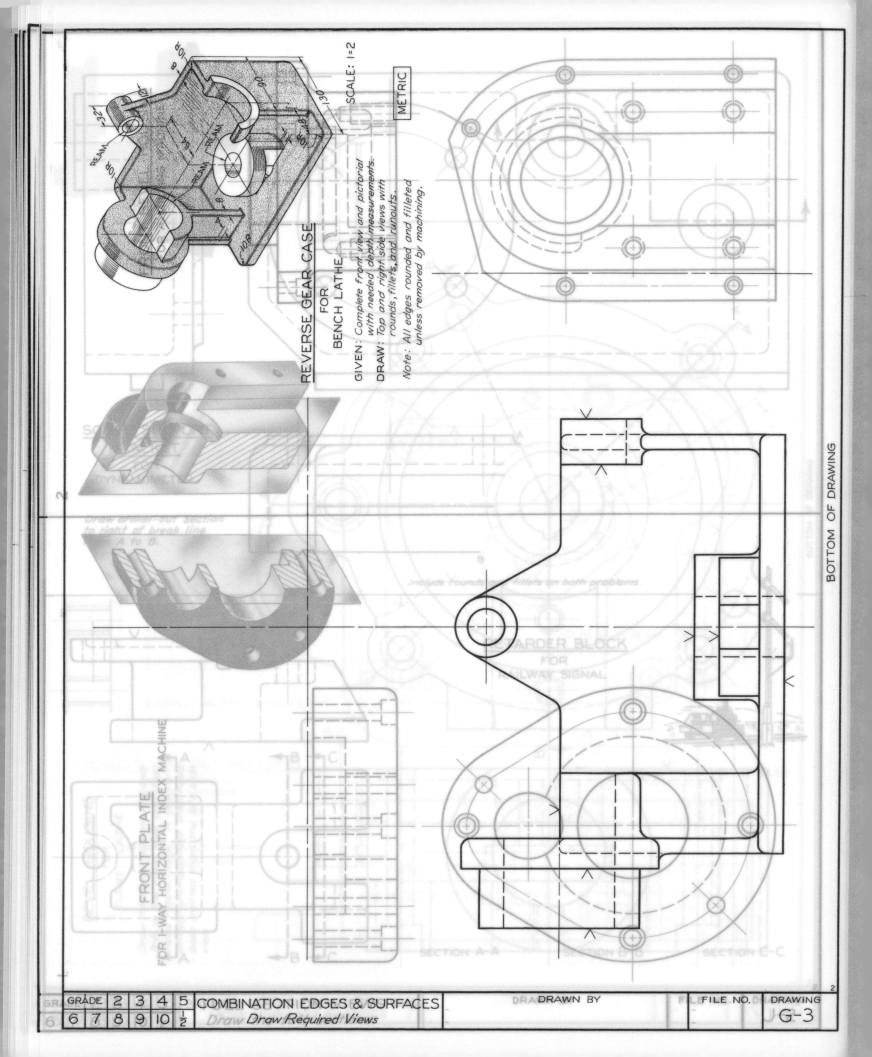

REVERSE GEAR CASE
FOR
BENCH LATHE

SCALE: 1=2

METRIC

GIVEN: *Complete front view and pictorial with needed depth measurements.*
DRAW: *Top and right side views with rounds, fillets, and runouts.*

Note: *All edges rounded and filleted unless removed by machining.*

RETARDER BLOCK
FOR
RAILWAY SIGNAL

Include rounds and fillets on both problems.

FRONT PLATE
FOR 1-WAY HORIZONTAL INDEX MACHINE

Draw broken-out section to right of break line from A to B.

SECTION A-A   SECTION B-B   SECTION C-C

GRADE	2	3	4	5	COMBINATION EDGES & SURFACES	DRAWN BY	FILE NO.	DRAWING
6	6	7	8	9	10	½	Draw Required Views			G-3

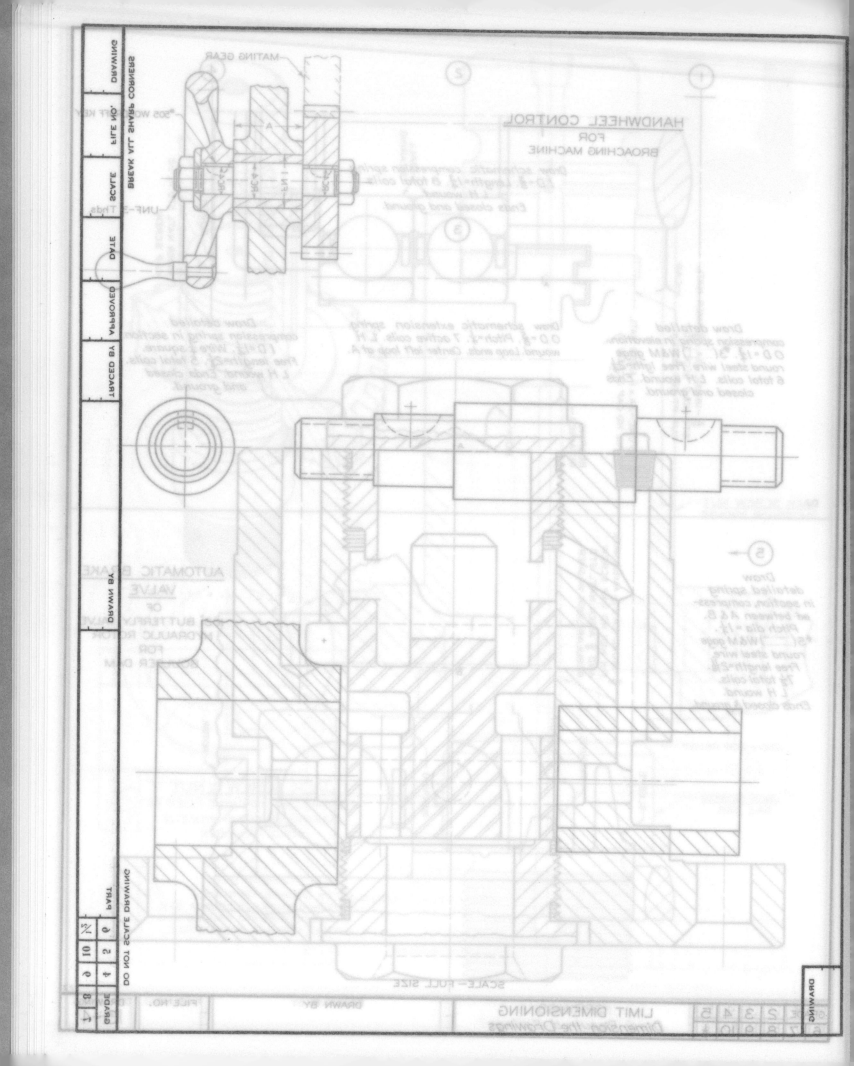

DRAWING

DO NOT SCALE DRAWING

BREAK ALL SHARP CORNERS

DRAWN BY

TRACED BY

APPROVED

DATE

SCALE

FILE NO.

DRAWING

PART

GRADE	4	5	6
	8	9	10
	7		½

BREAK ALL SHARP CORNERS

DRAWING	FILE NO.	SCALE	DATE	APPROVED	TRACED BY

DRAWN BY

DO NOT SCALE DRAWING

GRADE	PART			
8	7			
9	4			
10	5			
6	3/4			

DRAWING

DRAWING

GRADE	4	5	6	PART	DRAWN BY	TRACED BY	APPROVED	DATE	SCALE	FILE NO.	DRAWING
7	8	9	10	½							